I0025378

Blockchain

Psychology

Blockchain, Behavior & the Quest for a Better Future

CJ Carswell

All rights are reserved, and no part of this publication may be reproduced, distributed, or transmitted in any manner, whether through photocopying, recording, or any other electronic or mechanical methods, without the explicit prior written permission of the publisher. This restriction applies to any form or means of reproduction or distribution.

Exceptions to this rule include brief quotations that may be incorporated into critical reviews, as well as certain other noncommercial uses that are allowed by copyright law. Any such usage must adhere to the specified conditions and permissions outlined by the copyright holder.

BlocPsych, LLC

30 N Gould Street, Suite 28551

Sheridan, WY 82801

www.blocpsych.com

First Edition: October 2025

(Revised)

The publisher is not responsible for websites, or their content, that are not owned by the publisher.

ISBN: 979-8-9997084-2-7 (print edition)

Acknowledgements

I am deeply indebted to the insightful individuals who were early readers for this manuscript. Your thoughtful critiques and suggestions were invaluable in shaping the final narrative.

To the unwavering support system in my personal life, your understanding and encouragement provided the bedrock upon which this book was built. Your belief in this project gave me the motivation to give my best and I am deeply grateful for your love and support.

A Note on the Development of This Book

This book represents the culmination of research and conceptual development that began during my doctoral studies in business psychology. The core theoretical framework emerged from explorations within my coursework, sparking a deep interest in the intersection of blockchain technology and human behavior.

Given the expansive nature of blockchain technology and its potential implications for psychological principles, the initial scope of this research extended beyond the parameters of a traditional business psychology dissertation. Therefore, this book serves as an introduction of blockchain concepts to the field of psychology academia, particularly business psychology, where the human element within technological advancements is increasingly relevant. While a growing body of scholarly work explores the connection between blockchain and behavioral economics, this book aims to further bridge the gap and stimulate discussion regarding the broader psychological dimensions of this transformative technology.

In the process of developing this work, I employed several resources to aid in research, writing, and presentation:

AI Assistance: Gemini AI served as a research assistant, helping to identify and synthesize information across a broad range of sources. It also aided with drafting and editing portions of the manuscript along with the use of ProWritingAid AI for editing and content analysis. I have critically reviewed and refined all AI-generated content to ensure accuracy and alignment with the core arguments of this book.

External Review: I benefited greatly from the insights of individuals with expertise in both blockchain technology and behavioral science. Contractors from Fiverr, as well as colleagues, generously served as readers, providing valuable feedback and perspectives that have enriched the development of this theory.

It is important to note that this work is not a dissertation submitted in fulfillment of the requirements for my doctoral degree. Instead, it stands as an independent exploration and articulation of the Theory of Blockchain Psychology, driven by a desire to introduce this novel framework to the academic community and encourage further empirical investigation through academic channels. My intention is that this book will serve as a foundational text for future research exploring the intricate relationship between blockchain technology and the human psyche.

Introduction: Blockchain and Behavioral Science

In an era defined by rapid technological advancements and pressing global challenges, the convergence of blockchain technology and behavioral science offers a unique opportunity to reshape our world. Blockchain, with its decentralized and transparent nature, holds the potential to revolutionize industries and empower individuals. However, realizing its true promise requires looking beyond its technical capabilities to understand how it aligns with—and can be enhanced by—human behavior.

This book explores the fascinating intersection of blockchain and behavioral science, demonstrating how understanding human decision-making, motivations, and biases can unlock blockchain's transformative power for creating a more equitable and sustainable future. It is written for everyone interested in learning about blockchain, particularly those looking beyond cryptocurrency to grasp its wider potential. While crypto, NFTs, and DeFi are part of the landscape, the focus here is on the foundational opportunities blockchain presents and how understanding human behavior can help us build better, safer, and more effective applications. It is also intended for social scientists, offering a bridge to understanding the relevance of this technology to the human condition.

Specifically, this work introduces the theory of blockchain psychology – a framework that aims to deepen our understanding of how humans interact with and adopt these transformative decentralized systems. The foundations of this theory are based primarily on the work of Daniel Kahneman and Amos Tversky, pioneers in the field of behavioral economics, with broader areas of social science also incorporated into this theoretical framework.

Acknowledgements..3

Introduction: Blockchain and Behavioral Science............ 6

Chapter 1: Demystifying Blockchain 15

What is Blockchain? **15**

Blockchain's Impact...**18**

Blockchain and Web 3.0 ....**19**

Diving Deeper into Blockchain Types: Public, Private, and
Consortium..**20**
    Public Blockchains 20
    Private Blockchains... 21
    Consortium Blockchains 22

Decentralized Autonomous Organizations (DAOs): The
Promise and Perils of Distributed Governance **23**
    The Thorny Path to Decentralized Governance: Key Challenges 24
        Achieving Consensus Among Diverse Stakeholders 24
        Managing Conflicts and Disputes 25
        Ensuring Security Against Attacks........................ 26
        Addressing Potential Power Imbalances.............................27

Reaching Consensus: The Engine of Blockchain **28**
    Proof-of-Work (PoW).. 29
    Proof-of-Stake (PoS) 31
    Other Consensus Mechanisms.................................. 32
    Choosing a Consensus Mechanism: Trade-offs........................ 33

Scalability Challenges in Blockchain: Overcoming Growing
Pains..**34**
    Challenges: The Bottlenecks of Growth: 35
    Potential Solutions: A Multi-Layered Approach...................... 36
        Layer-1 Scaling Solutions (Improving the Base Protocol) 36
        Layer-2 Scaling Solutions (Building on Top of Layer-1) 37
        Scalability Trade-offs: No One-Size-Fits-All................................ 38

Navigating the Regulatory Maze: Blockchain and the Law..39

Impact on Blockchain Adoption and Innovation.........................39

Complex Regulatory Approaches Across Jurisdictions..............40

Beyond Cryptocurrencies: A Universe of Applications........42

Supply Chain Management...43

Voting Systems 44

Digital Identity Management...45

Healthcare..46

New Economic Models 47

Government and Public Sector 49

Intellectual Property...50

Charity and Nonprofits 51

Blockchain & Social Media...51

Examples of Blockchain Social Media Platforms.......................52

Chapter 2: From Ancient Principles to the Decentralized Web 54

The Historical Roots of Blockchain: Ancient Practices, Modern Solutions 54

Decentralized Record-Keeping: Echoes of Ancient Ledgers 55

Redundancy 55

The Redundancy of Clay Tablets..55

Blockchain's Evolution of Redundancy 56

Accountability 56

Medieval Guilds: Cross-Referencing for Accountability.........56

Blockchain's Automated Audit Trail.......................................57

Ledgers & Registries 57

Historical Land Registries: Community-Based Security 57

Blockchain's Global Digital Ledger...58

Cryptography...58

Ancient Origins of Secure Communication 58

Blockchain's Leverage of Modern Cryptography....................59

Consensus 60

Tribal Councils Consensus Through Deliberation 61

Quaker Consensus Seeking Unity in Agreement 61

Byzantine Fault Tolerance: Designing for Unreliable
Participants 62

**The Evolution of the Internet: From Web 1.0 to the Promise
and Perils of Web 3.0 .... 63**

Web 1.0 (Read-Only Web) 63

Web 2.0 (Social Web) 64

Web 3.0: The Ambition of a Decentralized and Semantic Web .. 64

Challenges in Realizing Web 3.0 65

Scalability and Performance 65

Usability and User Experience 65

Regulation and Governance 66

Interoperability 66

Security and Smart Contract Vulnerabilities 66

The Transition from Web 2.0 66

Fulfilling the Internet's Original Vision: Reclaiming
Decentralization and Empowerment 67

Data Ownership and Control 67

Open and Permissionless Platforms 68

Community-Driven Governance and Value Creation 68

What Does it Mean to Own the Internet? 68

The Genesis of Bitcoin: A Response to Crisis and a Vision for the
Future 70

The 2008 Financial Crisis: A Catalyst for Change 70

Satoshi Nakamoto's Vision 71

Key Innovations of the Bitcoin Whitepaper 71

Motivations Behind Bitcoin 72

Beyond Bitcoin: Charting the Evolution of Blockchain Technology
.... 73

The Rise of Ethereum and Smart Contracts 74

The Evolution of Consensus Mechanisms 75

Building Blocks of Blockchain: Exploring the Technological
Precursors 75

Byzantine Fault Tolerance: Achieving Consensus in a
Decentralized World 77

Growth of the Blockchain Ecosystem 78

Recognizing Key Contributors to Blockchain 78

Vitalik Buterin 79

Gavin Wood Architecting the Future of Blockchain 79

The Ethereum Foundation Supporting a Decentralized
Ecosystem.. 79

The Bitcoin Foundation Advocating for Bitcoin's Growth...... 80

Chapter 3: The Human Side of Blockchain 82

Understanding Our Biases in the Blockchain Era: The Cognitive
Landscape of Web3 85

Prospect Theory and Loss Aversion: Navigating Volatility
and Digital Wealth... 85

Availability Bias: The Shadow of Scams and the Glare of Hype
.... 88

Excessive Coherence and Confirmation Bias: Echo Chambers
and Unchallenged Narratives in Communities..................... 90

Other Relevant Biases 92

**The Theory of Blockchain Psychology: A Framework for
Human-Centered Design 94**

Underlying Assumptions.. 94

Core Components of Blockchain Psychology........................... 95

A Model for Understanding Blockchain Adoption and Use........ 97

Psychological Factors 98

Blockchain Characteristics... 98

User Outcomes 98

Predictive Power 99

Connecting Biases to System Noise...................................... 100

How specific biases contribute to noise 101

Behavioral Intervention for System Noise Solutions 102

Blockchain's Potential to Influence Biases............................. 103

Culture, Motivations, and User Decision-Making 105

Blockchain for Social Good: Leveraging Behavioral Insights 107

The Indispensable Human Element................................109

Who Uses Blockchain?... 112

Individuals 113

Businesses 114

Developers... 115

Communities 115

Chapter 4: Building a More Equitable & Sustainable World
.... 118

A Human-Centered Approach to Equity: Navigating Biases and Building Trust 118

Social Intelligence (Applying Behavioral Group Dynamics) 119

Emotional Intelligence (Applying Behavioral Principles of Trust and Motivation).. 120

Cultural Intelligence (Applying Behavioral Variation and Narrative Understanding) 121

Blockchain Applications for Equity: Behavioral Hurdles and Human-Centered Design 121

Financial Inclusion for the Underserved................................. 122

Land Rights and Property Management................................... 123

Supply Chain Transparency for Ethical Sourcing 124

Decentralized Governance (DAOs) for Community Empowerment
.... 126

Real-World Examples: Behavioral Insights in Practice 127

Blockchain & Sustainability: Behavioral Drivers for a Greener Future... 134

Blockchain's Environmental Impact: The Energy Debate 135

Energy-Efficient Alternatives (PoS and Beyond) 136

The Human Factor in the Energy Transition 137

Revolutionizing Resource Management & Traceability (Leveraging Blockchain for Environmental Good)................... 138

Tracking and Verifying Sustainability Initiatives (Building Trust and Credibility) 139

Incentivizing Green Behavior (Designing Effective Systems) 140

Navigating The Path to A Greener Blockchain: A Holistic Approach... 140

Chapter 5: The Convergence of Blockchain and AI 142

Enhancing AI Capabilities with Blockchain........................142

Enhancing Blockchain Capabilities with AI........................144

Protecting Digital Assets: Likeness and Intellectual Property 146

Use Cases 149

Challenges and Considerations 150

Towards a Human-Centered AI and Blockchain Future.....151

Chapter 6: The Future of Blockchain: Scenarios, Challenges, and Responsible Development.................. 155

Shaping Blockchain's Future 155
Scenario 1: Decentralized Trust and Individual Empowerment 156
Scenario 2: Community Empowerment and Economic Inclusion 159
Scenario 3: New Income Streams and Economic Models 163
Scenario 4: Navigating Challenges and Ethical Considerations: A Deeper Dive 167

Collaboration, Pragmatism, and Human-Centricity 176

Chapter 7: Navigating the Blockchain Information Landscape.. 177

Identifying Credible Information Sources 177
For General Information & Analysis 177
For a Deeper Dive.. 179

Beyond News Outlets: Research and Community 179

Fact-Checking and Evaluating Information.......................181
Tips for Evaluating Information... 181
Detecting Misinformation... 183

Understanding Blockchain Performance.........................**184**

Credible Reports and Resources on Blockchain Performance.. 184

Resources for Blockchain Performance Analysis Tools............. 186

Key Factors to Consider When Evaluating Blockchain
Performance... 186

Chapter 8: Glossary of Key Terms ***189***

References ***209***

About the Author.. ***228***

Chapter 1: Demystifying Blockchain

Blockchain technology is rapidly transforming our world, but its core principles are often obscured by hype and technical jargon. This chapter aims to demystify blockchain, explaining its key components and diverse applications for those unfamiliar with the technology. By building a solid foundational understanding of what blockchain is, what it does, and how it works, we can better assess its true value and potential for the future. Let's dive in.

What is Blockchain?

At its heart, blockchain is more than just a piece of software or a database. It's a foundational distributed ledger technology that functions as a secure and transparent system for managing and recording data across a network of computers (Di Ciccio, 2024). Imagine it as a shared, continuously updated digital spreadsheet or ledger that is duplicated and spread across many different computers (nodes) around the world, rather than being stored in one central location.

At its core, a blockchain comprises several key components working in concert:

Decentralized Ledger

Unlike traditional databases controlled by a single entity (like a bank or a company), the blockchain maintains a shared, distributed ledger. Every participant (node) on the network holds a copy of this continuously updated record. This decentralization is fundamental, ensuring transparency and eliminating single points of failure or control (Vendette & Thundiyil, 2023).

Cryptographic Security

Robust cryptographic techniques are employed to secure transactions and data. Each transaction is bundled together with others into a "block." This block is then cryptographically linked to the previous block using a unique digital fingerprint called a hash (Ahmed, 2023). This hashing and linking mechanism creates a chronological chain of blocks (hence "blockchain") and makes the data within them tamper-proof. Any attempt to alter a past block would change its hash, breaking the link to the next block and immediately alerting the network to the attempted manipulation.

Consensus Mechanisms

To validate new transactions and ensure that all the distributed nodes agree on the correct state of the blockchain, networks utilize various consensus mechanisms. These are automated protocols or algorithms (like Proof-of-Work or Proof-of-Stake) that enable participants to collectively agree on the legitimacy of transactions before they are added to the chain (Conway, 2022). This collective agreement process fosters trust within the network without the need for a central intermediary or authority (OSL, 2025a).

Diverse Applications

While initially conceived as the underlying technology for cryptocurrencies like Bitcoin, the capabilities of blockchain extend far beyond financial transactions. Its secure, transparent, and decentralized nature makes it applicable across a wide range of industries and use cases, from tracking goods in a supply chain to verifying digital identities and enabling new forms of digital ownership (Bialas, 2024).

The interplay of these core components results in a system with several defining characteristics:

Decentralization

Information is not stored on a single server but is distributed across numerous computers (nodes), each holding a copy of the ledger. This distribution ensures data integrity and eliminates reliance on a central authority, making the system more resistant to censorship and single points of failure (Vendette & Thundiyil, 2023).

Immutability

Once a transaction is verified and recorded in a block, and that block is added to the chain, it becomes extremely difficult, practically impossible under normal network conditions, to alter or delete it. This permanence creates an auditable and chronological history of all transactions, fostering trust and transparency (Ahmed, 2023).

Transparency

While the underlying technology ensures the immutability and security of records, most public blockchains are designed to be transparent in that all participants can view the transaction history of the ledger. However, user identity is typically protected through cryptographic techniques, ensuring a degree of pseudonymity rather than full anonymity (McLeod, n.d.).

Blockchain's Impact

Building on these core characteristics, blockchain technology empowers individuals and communities in several significant ways, challenging traditional centralized models:

Decentralized Governance

Blockchain enables the creation of Decentralized Autonomous Organizations (DAOs), which can foster more democratic and transparent decision-making within online and offline communities by allowing token holders to vote on proposals (Santana & Albareda, 2024).

Enhanced Transparency and Accountability

Blockchain's public or shared ledger creates a transparent and auditable record of actions and transactions. This promotes trust and accountability in various contexts, including community management, business operations, and tracking the flow of funds (Bialas, 2024).

Tokenized Incentives

Blockchain-based tokens can be used to create incentive structures that reward positive contributions or desired behaviors within communities or platforms, encouraging engagement and the creation of valuable content or services (Santana & Albareda, 2024; Wei et al., 2024).

Data Privacy and Security

While public blockchains are transparent, blockchain can enhance online privacy and security by giving users greater control over their own data and digital identities through self-sovereign identity solutions (Omar et al., 2025.

Censorship Resistance

Due to its decentralized nature, where data is distributed across many nodes, blockchain networks are inherently more resistant to censorship or control by any single entity compared to centralized platforms (GeeksforGeeks, 2023; Vendette & Thundiyil, 2023).

Blockchain and Web 3.0

These impacts are central to the vision of Web 3.0, often referred to as the decentralized web. Building on the static Web 1.0 and the interactive but centralized Web 2.0, Web 3.0 envisions an internet where users have greater control over their data, identity, and online interactions (Minhaz et al., 2024). Blockchain is a foundational technology of Web 3.0, providing the transparent, secure, and tamper-proof infrastructure for decentralized applications (dApps) and services where users can interact directly with each other and dApps, bypassing intermediaries for a potentially more equitable online experience.

Diving Deeper into Blockchain Types: Public, Private, and Consortium

While the fundamental principles of blockchain provide these benefits, its implementation varies depending on the intended use case and required level of access and control. In the world of blockchain, one size does not fit all. It's essential to delve deeper into the diverse landscape of blockchain types to understand how the technology is being tailored for specific needs.

Public Blockchains

Permissionless and Open: Public blockchains are "permissionless," meaning anyone can participate in the network. Anyone can read the ledger, submit transactions, and participate in the consensus process (e.g., by mining or validating) (Sharma, 2024).

Characteristics: These blockchains are typically highly decentralized, with no single entity controlling the network. This fosters maximum transparency as all transactions are publicly viewable, though user identity is usually pseudonymous (McLeod, n.d.). Security relies heavily on cryptography and the collective effort of the decentralized network (Ahmed, 2023).

Use Cases: Ideal for applications requiring maximum transparency, decentralization, and censorship resistance, such as cryptocurrencies (Bitcoin, Ethereum), decentralized finance (DeFi), and public digital registries.

Potential Counterargument: While aiming for maximum decentralization, some argue that the concentration of mining power (in PoW) or staking power (in PoS) can introduce new forms of centralization in practice (Brookings Institution, 2025).

Private Blockchains

Permissioned and Controlled: Private blockchains are "permissioned" networks, where access and participation are restricted and controlled by a central authority or organization (Sharma, 2024).

Characteristics: A single entity typically governs the network, controls who can participate, and defines their permissions. This offers greater privacy as transactions are only visible to authorized participants, making them suitable for use cases where confidentiality is crucial. They can also offer faster transaction speeds and higher throughput due to the limited number of known participants.

Use Cases: Suitable for internal organizational use or applications requiring high privacy and efficiency within a controlled environment, such as supply chain management within a single company, internal voting systems, or managing sensitive private data.

Consortium Blockchains

Hybrid Approach: Consortium blockchains combine elements of both public and private blockchains. They are permissioned networks governed by a group of pre-selected organizations or institutions, rather than a single entity (Sharma, 2024).

Characteristics: Multiple organizations collaborate to validate transactions and maintain the blockchain, promoting shared governance and responsibility. Access is granted to specific participants within the consortium, while certain data may remain private to specific groups.

Use Cases: Ideal for inter-organizational collaboration where trust is needed among a known group, such as supply chain networks involving multiple companies, cross-border payments between banks, or data sharing between healthcare providers.

Understanding the different types of blockchains and their characteristics helps individuals and organizations make informed decisions about which blockchain solution best suits their specific needs and the trade-offs involved. This choice often depends on balancing the desire for decentralization and openness with the requirements for privacy, control, and performance.

Decentralized Autonomous Organizations (DAOs): The Promise and Perils of Distributed Governance

Building on the concept of decentralized governance enabled by blockchain, Decentralized Autonomous Organizations (DAOs) represent an innovative approach to community governance. Envisioning member-owned entities operating without traditional centralized leadership, DAOs aim to democratize decision-making processes for managing shared resources, funding projects, and enacting collective choices through code in the form of smart contracts (Santana & Albareda, 2024). Examples like Friends With Benefits (FWB) for community building, PleasrDAO for collective asset management, and LexDAO for legal engineering in Web3 showcase the diverse applications of this model (Debutinfotech, 2025).

However, while the promise of democratic and decentralized governance is compelling, the practical realities of creating and managing effective DAOs are fraught with complexities and challenges that often go unacknowledged in introductory discussions. Achieving true decentralization in governance is not a simple technological feat; it involves navigating intricate human dynamics and addressing potential pitfalls inherent in collective decision-making (Fan, 2024).

The Thorny Path to Decentralized Governance: Key Challenges

Achieving Consensus Among Diverse Stakeholders

DAOs often comprise individuals with varying levels of expertise, motivations, and vested interests. Reaching consensus on critical decisions within such diverse groups can be a slow, cumbersome, and sometimes contentious process. Different stakeholders may have conflicting priorities, leading to gridlock or the dominance of certain vocal minorities or large token holders. The very nature of decentralized systems, lacking a central authority to impose decisions, necessitates robust and well-defined mechanisms for proposal submission, discussion, and voting. The absence of clear processes or the presence of low participation rates can undermine the legitimacy and efficiency of DAO governance (Fan, 2024).

Real-world Example

Early DAOs, like "The DAO" (which suffered a major hack), struggled with establishing clear decision-making processes and resolving disagreements on critical proposals. The lack of a well-defined mechanism to address unforeseen vulnerabilities ultimately led to its downfall (ImmuneBytes, 2023).

Potential Solutions and Challenges

Implementing clear voting thresholds or utilizing quadratic voting mechanisms (where voting power increases less than proportionally to the amount of tokens held) can help mitigate the influence of large holders. Establishing working groups and fostering robust discussion forums can facilitate consensus. However, ensuring broad, informed participation remains a significant challenge, often requiring active community management and education (Fan, 2024).

Managing Conflicts and Disputes

Disagreements and conflicts are inevitable in any human collective, and DAOs are no exception. Without a traditional hierarchical structure for dispute resolution, DAOs must develop their own mechanisms for addressing conflicts fairly and efficiently. This can range from internal mediation processes facilitated by elected community members to more formal on-chain dispute resolution systems. The lack of clear and trusted dispute resolution mechanisms can lead to community fragmentation and hinder the DAO's ability to function effectively (Fan, 2024).

Real-world Example

In various NFT-focused DAOs, disagreements over the direction of the community or the management of the treasury have led to heated debates, and in some cases, the splintering of the community (Debutinfotech, 2025).

Potential Solutions and Challenges

Establishing clear codes of conduct and implementing internal mediation or arbitration processes can help. Exploring the use of decentralized dispute resolution platforms that leverage external expertise or on-chain voting can provide pathways for managing conflicts. However, enforcing these resolutions in a truly decentralized and potentially pseudonymous environment presents practical difficulties (Fan, 2024).

Ensuring Security Against Attacks

While the underlying blockchain technology offers a degree of security, DAOs themselves, particularly their smart contract logic and governance mechanisms, can be vulnerable to attacks and manipulation. Exploits of governance proposals, Sybil attacks (where a single entity controls multiple voting identities), and "whale" manipulation (where large token holders exert undue influence) are significant security concerns (Fan, 2024). The decentralized nature of DAOs can sometimes make it slower and more challenging to respond to and mitigate security threats compared to centralized organizations.

Real-world Example

Several DAOs have been targeted by governance attacks where malicious actors have accumulated enough voting power to pass harmful proposals, such as transferring treasury funds to their control (ImmuneBytes, 2024).

Potential Solutions and Challenges

Implementing robust smart contract audits, employing multi-signature wallets for treasury management, introducing time-lock mechanisms for critical governance changes, and continuously monitoring for suspicious voting patterns can enhance security. However, no system is entirely foolproof, and the immutability of blockchain can make recovering from an exploit difficult (ImmuneBytes, 2023).

Addressing Potential Power Imbalances

Despite the ideal of democratic governance, power imbalances can still emerge within DAOs. Early adopters, large token holders, or technically savvy members can disproportionately influence decision-making processes. This can lead to the marginalization of smaller stakeholders and undermine the principles of true decentralization. Ensuring equitable participation and representation remains a significant challenge (Fan, 2024).

Real-world Example

In some early DAOs with simple token-weighted voting, a small number of large token holders effectively controlled all governance decisions, leading to concerns about centralization in disguise (Brookings Institution, 2025).

Potential Solutions and Challenges

Exploring alternative voting mechanisms like quadratic voting or implementing delegation systems can help mitigate power imbalances. However, these mechanisms can introduce their own complexities and may not fully eliminate the influence of wealth or technical expertise (Fan, 2024).

While Decentralized Autonomous Organizations offer a compelling vision for the future of community governance and resource management, their successful implementation requires a deep understanding of the inherent challenges and complexities involved. Addressing issues such as achieving consensus, managing disputes, ensuring security, and mitigating power imbalances is crucial for realizing the full potential of DAOs as truly democratic and resilient organizations. The evolution of best practices, the development of robust governance frameworks, and a continued focus on user-centric design will be essential for navigating these challenges and fostering the growth of effective and equitable DAOs within the Web3 ecosystem.

Reaching Consensus: The Engine of Blockchain

Building on the understanding of blockchain types and decentralized governance, a fundamental question remains: how do all the distributed nodes in a decentralized network agree on the validity of transactions and the exact state of the ledger without a central authority? This is where consensus mechanisms come into play. They are the fundamental protocols and processes that allow distributed nodes across the network to collectively agree on the next block of transactions to be added to the chain, ensuring the integrity, security, and immutability of the blockchain without relying on a central arbiter (OSL, 2025a). Understanding these mechanisms is key to appreciating the diverse capabilities and trade-offs inherent in different blockchain technologies.

Proof-of-Work (PoW)

The Work is the Proof

In PoW, nodes, known as "miners," engage in a computationally intensive competition to solve a complex cryptographic puzzle. This puzzle involves finding a specific number (a "nonce") such that when combined with the data in the block and run through a hash function, it produces a hash output that meets a certain target requirement (e.g., starts with a specific number of zeros) (OSL, 2025a). This process is essentially trial-and-error, requiring significant computational power and energy.

Mining Process

Miners repeatedly guess nonces and calculate the hash until one finds a valid hash below the target. The difficulty of the puzzle is adjusted by the network to control the rate at which new blocks are found (e.g., roughly every 10 minutes for Bitcoin) (OSL, 2025a). The first miner to find a valid hash "wins" the right to add the next block of validated transactions to the blockchain and is rewarded with newly minted cryptocurrency and transaction fees.

Security through Cost

The security of a PoW network lies in the immense computational effort and energy cost required to find a valid block hash. To alter a transaction or create a fraudulent block, an attacker would need to redo the work for that block and all subsequent blocks faster than the rest of the network combined. This is known as a 51% attack (OSL, 2025b). The economic and energy cost of acquiring and running enough computing power to achieve over 50% of the network's total hash rate makes such an attack prohibitively expensive for large, established PoW chains like Bitcoin (OSL, 2025a).

Advantages

High Security (proven against 51% attacks on large networks), Proven Decentralization (permissionless participation), Battle-Tested (used by Bitcoin for over a decade), Clear Incentives (block rewards and fees) (OSL, 2025a, OSL, 2025b).

Drawbacks

Environmental Concerns (enormous energy consumption), Scalability Issues (limited transactions per second, slower speeds, higher fees), Hardware Centralization Risk (concentration of mining power in pools/regions) (OSL, 2025a).

Real-World Examples

- Bitcoin (BTC)
- Litecoin (LTC)
- Dogecoin (DOGE)

Proof-of-Stake (PoS)

Staking for Validation

In PoS, nodes, called "validators," do not compete by solving computational puzzles. Instead, they "stake" a certain amount of the network's cryptocurrency as collateral to participate in the consensus process. This stake acts as a financial bond and an incentive to act honestly (Conway, 2022; OSL, 2025a).

Validator Selection

Validators are selected to create and validate new blocks based on a combination of factors, typically including the amount of cryptocurrency they have staked, the length of time their stake has been active, and often an element of randomness to prevent predictability and centralization (OSL, 2025a).

Security through Incentives

The security of a PoS network relies on economic incentives. Validators are rewarded for acting honestly (e.g., with transaction fees or a small amount of newly minted crypto). Conversely, if a validator behaves maliciously (e.g., attempts to validate fraudulent transactions or create competing chain histories), their staked cryptocurrency can be partially or entirely confiscated by the protocol – a process called "slashing." This financial penalty makes malicious behavior economically irrational, aligning the validator's self-interest with the network's security (OSL, 2025a).

Advantages

Energy Efficiency (drastically lower consumption than PoW), Improved Scalability (faster block times, higher throughput potential), Economic Security (attack cost is acquiring and losing stake), Lower Barrier to Entry (for validation compared to PoW hardware), Reduced Hardware Centralization Risk (OSL, 2025a).

Drawbacks

"Nothing at Stake" Problem (mitigated by slashing), Wealth Concentration Risk (larger stakes have higher chance of reward), Potential for Initial Centralization (based on initial token distribution), Complexity (designing robust protocols), Relative Novelty (less long-term testing than PoW) (OSL, 2025a).

Real-World Examples

- Ethereum (ETH) (after transitioning from PoW)
- Cardano (ADA), Polkadot (DOT)
- Avalanche (AVAX)

Other Consensus Mechanisms

Beyond the two dominant mechanisms, many other variations and entirely different approaches exist, often designed for specific use cases or to address the trade-offs inherent in PoW and PoS (Conway, 2022; OSL, 2025a).

Delegated Proof-of-Stake (DPoS)

Token holders vote for a limited number of "delegates" who validate transactions on their behalf, offering faster speeds but increased centralization (OSL, 2025a; Roy, 2023). (Examples: EOS, Tron).

Practical Byzantine Fault Tolerance (PBFT) and Variants

Nodes communicate directly to agree, suitable for permissioned networks requiring high speed and finality but not for large public chains (OSL, 2025a). (Examples: Hyperledger Fabric, Ripple).

Proof-of-Authority (PoA)

Relies on pre-selected, reputable validators, offering high speed and efficiency but high centralization (GeeksforGeeks, 2025; OSL, 2025a). (Examples: VeChain, BNB Chain).

Choosing a Consensus Mechanism: Trade-offs

The selection of a consensus mechanism is critical, fundamentally impacting the network's characteristics based on inherent trade-offs. The "best" mechanism depends entirely on the intended use case (Conway, 2022; OSL, 2025a). Key trade-offs include:

Decentralization vs. Performance

Maximizing decentralization often limits speed (PoW), while increasing performance may require more centralization (DPoS, PBFT).

Security vs. Efficiency

PoW is secure but energy-intensive. PoS is more efficient but relies on economic security and has different risks. Permissioned BFT/PoA are highly efficient but rely on pre-established trust or permissioning rather than open, economic security.

Openness vs. Control

Permissionless chains (PoW, PoS) are open but harder to control. Permissioned chains (PBFT, PoA) offer control but restrict participation.

Finality

Some offer immediate finality (PBFT), while others have probabilistic finality (PoW, PoS).

Consensus mechanisms are the backbone of blockchain technology, defining how trust is established and maintained in a distributed environment. By understanding the different mechanisms, their underlying principles, advantages, disadvantages, and the trade-offs they embody, we gain a deeper appreciation for the diverse landscape of blockchain projects and their potential applications. The ongoing evolution of consensus mechanisms continues to push the boundaries of what decentralized technologies can achieve (OSL, 2025a).

Scalability Challenges in Blockchain: Overcoming Growing Pains

Moving from how transactions are agreed upon, we must consider the practical limitations as network activity grows. As blockchain adoption increases, many networks face significant scalability challenges – the ability to handle a growing number of transactions efficiently without compromising speed or increasing costs (Zhou et al., 2023). It's a critical hurdle for widespread adoption, as real-world applications require throughput comparable to existing centralized systems.

Challenges: The Bottlenecks of Growth

The core scalability challenge in decentralized blockchains stems from the fundamental requirement that every full node must process and validate every transaction to maintain security and consensus. This inherent design, while crucial for decentralization and trustlessness, creates bottlenecks:

Transaction Throughput Limitations

Blockchains can only process a limited number of transactions per second (TPS). This limit is constrained by factors such as block size, block interval, and the efficiency of the consensus mechanism (Zhou et al., 2023). When transaction volume exceeds capacity, a backlog forms.

High Transaction Fees

As the network becomes congested, users compete for limited block space, driving up transaction fees (e.g., "gas fees" on Ethereum) (Zhou et al., 2023). High fees make micro-transactions or frequent interactions economically unfeasible.

Network Congestion and Latency

High traffic leads to delays in transaction confirmation, resulting in poor user experience and hindering applications requiring fast finality (Zhou et al., 2023).

Storage Capacity Constraints

Every full node must store a complete copy of the blockchain's history. As transactions grow, the blockchain size increases, demanding more storage from nodes (Zhou et al., 2023). If requirements become too high, it can price out individual users or smaller entities from running full nodes, potentially leading to centralization.

Computation Load

Beyond storage, nodes must validate transactions and execute smart contracts, increasing computational demands with complexity and volume (Zhou et al., 2023).

Potential Solutions: A Multi-Layered Approach

Addressing blockchain scalability is an active area of research and development, involving various strategies categorized into Layer-1 and Layer-2 solutions, along with other complementary approaches (Zhou et al., 2023).

Layer-1 Scaling Solutions (Improving the Base Protocol)

These improve processing directly on the main chain.

Consensus Mechanism Optimization

PoS, DPoS, PBFT achieve faster block times and higher throughput than PoW (OSL, 2025a; Zhou et al., 2023). (Example: Ethereum's PoS transition).

Block Size and Block Interval Adjustments

Increasing block size or decreasing block time can increase throughput but involve trade-offs with decentralization, security, and storage (Zhou et al., 2023).

Sharding

Dividing the network state and processing load into smaller "shards" enables parallel processing to increase throughput (Zhou et al., 2023). (Example: Ethereum 2.0's long-term vision).

Layer-2 Scaling Solutions (Building on Top of Layer-1)

These move processing off-chain, using Layer-1 for final settlement and security (Zhou et al., 2023).

General Concept: Off-Chain Processing

Transactions occur on a secondary layer, enforced by Layer-1 smart contracts (Zhou et al., 2023).

State Channels

Allow direct off-chain transactions between participants with final settlement on Layer-1 (Zhou et al., 2023). (Example: Lightning Network for Bitcoin).

Rollups (Optimistic and Zk-)

Bundle off-chain transactions and post summaries/proofs on Layer-1. Optimistic assume validity with fraud proofs; Zk-Rollups use validity proofs (Zhou et al., 2023). (Examples: Optimism, Arbitrum (Optimistic); zkSync, StarkNet (Zk)).

Sidechains

Independent blockchains linked to the main chain via a two-way peg (Zhou et al., 2023). (Example: Polygon).

Off-Chain Data Storage / Data Availability Layers

Store data off-chain but ensure availability, posting commitments on Layer-1 (Zhou et al., 2023).

Scalability Trade-offs: No One-Size-Fits-All

Choosing scalability solutions involves navigating trade-offs: Decentralization vs. Performance, Security vs. Efficiency/Cost, Complexity vs. Usability, Generality vs. Specialization (Zhou et al., 2023). Different solutions prioritize different aspects based on the use case.

Real-World Examples and Ongoing Efforts

Projects pursue various strategies:

- Bitcoin focuses on Layer-2 (Lightning).
- Ethereum uses a multi-pronged approach (PoS + Sharding + Rollups).

- Solana uses a unique Layer-1 approach (PoH + PoS).
- Polkadot/Cosmos focus on interoperability (OSL, 2025a; Zhou et al., 2023).

The pursuit of blockchain scalability is a dynamic field. No single solution is a silver bullet; the future likely involves a combination of approaches tailored to specific network needs (Zhou et al., 2023).

Navigating the Regulatory Maze: Blockchain and the Law

Beyond technical and scalability challenges, blockchain technology operates in a rapidly evolving regulatory landscape that presents significant complexities and uncertainties (Narain & Moretti, 2022). Navigating this space requires understanding the diverse, often conflicting, approaches taken globally and the specific challenges this poses for blockchain projects.

Impact on Blockchain Adoption and Innovation

The regulatory landscape profoundly impacts adoption. Clear regulations foster confidence and innovation, while unclear or restrictive rules hinder investment, stifle development, create barriers, fragment markets, and encourage regulatory arbitrage (Narain & Moretti, 2022).

Complex Regulatory Approaches Across Jurisdictions

There is no single global approach. Strategies vary from strict prohibition to cautious innovation-friendly to wait-and-see (Narain & Moretti, 2022). These differences manifest in:

Legal Classification

Varying definitions of crypto assets (security, commodity, property, utility token) with different legal implications (e.g., US SEC vs. CFTC, EU MiCA framework, Switzerland's approach) (Narain & Moretti, 2022).

Taxation

Diverse approaches to taxing crypto (capital gains, income, property) creating complexity for users and businesses (Narain & Moretti, 2022).

Licensing and Registration

Varying requirements for crypto businesses (exchanges, wallets) across countries (Narain & Moretti, 2022).

AML/CFT

Implementation of FATF standards differs, challenging decentralized contexts (Narain & Moretti, 2022).

Consumer and Investor Protection

Rules vary regarding disclosures, suitability, and handling fraud (Narain & Moretti, 2022).

Securities Regulation

Application of securities laws (like the US Howey Test) to token offerings is a major focus (Narain & Moretti, 2022).

Specific Challenges for Blockchain Projects

Projects face a high compliance burden across jurisdictions, uncertainty in new areas (DeFi, DAOs, NFTs), difficulty operating globally, risk of sudden changes, and challenges defining responsibility in decentralized systems (Narain & Moretti, 2022).

Addressing the Challenges: Towards Clarity and Collaboration

Developing effective regulations requires collaboration between industry and regulators, international standards, sandboxes, and education (Narain & Moretti, 2022). Key entities involved include National Governments, International Organizations (FATF, FSB, IOSCO), Industry Associations, and Standards Bodies (ISO) (Narain & Moretti, 2022).

Blockchain's regulatory landscape is characterized by significant uncertainty and complexity, stemming from varying legal classifications, differing national approaches to key issues like taxation and licensing, and the lack of clear frameworks for novel decentralized applications. This complexity poses substantial challenges but also driving efforts towards greater clarity and international coordination to foster responsible innovation (Narain & Moretti, 2022).

Beyond Cryptocurrencies: A Universe of Applications

While cryptocurrencies like Bitcoin brought blockchain into the spotlight, its applications extend far beyond the financial sector. Blockchain's core properties—decentralization, immutability, and transparency—make it a foundational technology applicable across numerous industries to enhance trust, efficiency, and security (Bialas, 2024). In this section, we introduce some key areas where blockchain is being applied, providing a basic understanding of what these applications are and their fundamental benefits and challenges.

Supply Chain Management

Applying blockchain to track goods from origin to consumer for increased transparency, traceability, and efficiency (Bialas, 2024).

Examples: Walmart & IBM Food Trust

Supply Chain Management		
	Walmart's Food Traceability Program	**IBM Food Trust**
Focus	Tracking food products from farm to table to improve food safety and transparency.	A broader platform for enhancing food safety, traceability, and efficiency across the food supply chain.
Benefits	Faster Traceability Improved Accountability Reduces Waste Increased Consumer Trust	Enhanced Food Safety Reduced Fraud Improved Efficiency Increased Sustainability Greater Consumer Trust
Key Features	Immutable record of food product journey. Transparency for stakeholders.	Verification of Certifications. Secure Data Sharing Immutability
Connection	Walmart leverages the IBM Food Trust platform for its traceability initiative.	IBM Food Trust provides the underlying platform for Walmart's initiative.

Voting Systems

Applying blockchain to record votes for enhanced security, transparency, and accessibility (Bialas, 2024). Challenges include anonymity vs. preventing double voting, security of interfaces, and building public trust (Bialas, 2024).

Examples: Voatz & Follow My Vote

Voting Systems		
	Voatz	**Follow My Vote**
Focus	Mobile Voting with biometric verification.	Open-source voting software with emphasis on transparency and auditability.
Key Features	Biometric verification (Facial recognition, fingerprint scanning) Blockchain based audit trail Remote accessibility	Transparency (voters can verify ballots) Auditability Decentralization
Benefits	Increased accessibility Potential for increased voter turnout	Improved election integrity Increased voter participation
Challenges	Security Concerns Limited adoption due to lack of trust and independent audits.	Still in development Needs wider adoption

Digital Identity Management

Using blockchain to create a secure, verifiable, and user-controlled digital identity (Omar et al., 2025). Benefits include enhanced privacy and reduced reliance on central authorities (Omar et al., 2025). Challenges include interoperability and user responsibility (Omar et al., 2025).

Examples: Civic & Sovrin

Digital Identity Management		
	Civic	**Sovrin**
Focus	Secure and user-friendly identity verification solutions for businesses and individuals.	Decentralized, global identity network with self-sovereign identity for individuals.
Identity Storage	Encrypted and stored on a user's device.	Collection of verifiable credentials issued by trusted entities.
Key Features	Secure Identity Storage	Decentralization
	Selective Disclosure	Self-Sovereignty
	Reusability	Interoperability
	Blockchain-based verifiability	Privacy (zero-knowledge proofs)
Use Cases	KYC Compliance	Digital identity for all.
	Age Verification	Data Privacy
	Account Recovery	Secure Access
	Secure Access	Cross-border Identity

Healthcare

Applying blockchain for secure storage and sharing of patient records, drug traceability, and clinical trials (Bialas, 2024). Benefits include improved data security and interoperability (Bialas, 2024). Challenges include regulatory compliance and integrating with legacy systems (Bialas, 2024).

Examples: Medicalchain & BurstIQ

Healthcare		
	Medicalchain	**BurstIQ**
Focus	Patient-centric control and sharing of medical records.	Global health data network for various stakeholders.
Data Management	Patients control access to their medical records Immutable audit trail on the blockchain.	LifeGraph-comprehensive, secure record of individual health data shared with permissioned parties.
Key Features	Patient-centric Secure data sharing. Interoperability Data Integrity	Data ownership and control. Secure data exchange. Interoperability Advanced analytics
Use Cases	EHR Management Telemedicine Clinical Trials Medical Research	Population Health Management Personalized Medicine Value-based Care

New Economic Models

The emergence of DeFi, creator economies, and decentralized sharing economies built on blockchain (Bialas, 2024). Benefits include increased accessibility and new opportunities for value creation (Bialas, 2024). Challenges include regulatory uncertainty and security risks (Bialas, 2024).

Examples: MakerDAO, Compound, Audius

New Economic Models			
	MakerDAO (DeFi)	**Compound (DeFi)**	**Audius (Music Streaming)**
Focus	Decentralized Stablecoin (DAI)	Decentralized lending and borrowing platform.	Decentralized music streaming platform.
Economic Model	Collateralization Stability fees Governance Tokens (MKR)	Algorithmic interest rate model.	Governance token (AUDIO), for governance, staking and rewards. Artist and fan rewards based on engagement.
Key Features	Creates DAI stablecoin. Users earn interest on collateral. Users participate in governance.	Users supply or borrow assets. Algorithmic interest rates based on supply and demand.	Direct artist-fan connection. Fairer revenue distribution. Community Ownership

Benefits	A more open and accessible financial system. Transparency & accountability	Increased efficiency and accessibility in lending and borrowing.	Empowers artists and fans. Creates a more equitable music industry.

Government and Public Sector

Applying blockchain for land registries, citizen digital identity, and transparent public spending (Bialas, 2024). Benefits include reduced fraud and increased efficiency (Bialas, 2024). Challenges include implementation complexity and political resistance (Bialas, 2024).

Examples: Estonia's e-Residency Program & Dubai's Blockchain Strategy

Government & Public Sector		
	Estonia's e-Residency Program	**Dubai's Blockchain Strategy**
Focus	Providing a transnational digital identity and access to Estonian digital services for global citizens.	Transforming Dubai into a leading global hub for blockchain technology across various sectors.
Blockchain Use	Securing digital identities and transactions.	Streamlining government processes, creating new industries, and establishing international leadership.
Key Features	Digital ID card Company formation and management. Digital signing Online banking Tax declaration Access to EU business environment	Dubai blockchain platform Smart Dubai initiative Dubai Future Accelerators Global Blockchain Challenge

Intellectual Property

Using blockchain to timestamp, register, and track ownership and usage of creative works (Bialas, 2024). Benefits include verifiable proof of ownership and immutable provenance (Bialas, 2024). Challenges include legal recognition and enforcement (Bialas, 2024).

Examples: Binded & Ascribe*

Intellectual Property		
	Binded (now Pixy)	**Ascribe**
Focus	Copyright registration and protection	Broader IP management, including copyright and rights transfer.
Blockchain Use	Bitcoin blockchain	Bespoke blockchain based system
Key Features	Ease of use Affordability Blockchain verification Copyright monitoring	Authorship claims Rights transfer Limited editions Provenance tracking

As of the publishing of this book, we learned that Ascribe is no longer operational.

Charity and Nonprofits

Applying blockchain to track donations and verify fund usage (Bialas, 2024). Benefits include increased donor trust and reduced fraud (Bialas, 2024). Challenges include onboarding and managing volatility (Bialas, 2024).

Examples: Give Track & BitGive Foundation.

Charity & Non-Profits		
	GiveTrack	**BitGive Foundation**
Focus	Real-time transparency and accountability for charitable donations.	Leveraging blockchain to improve philanthropy across various causes.
Blockchain Use	Bespoke blockchain-based platform.	Uses Bitcoin and other cryptocurrencies to facilitate donations and grants.
Key Features	Transparency for donors Accountability for nonprofits Efficiency in the donation process Increased trust	GiveTrack platform Promotes Bitcoin for Charity Provides disaster relief Partners with nonprofits to implement blockchain solutions

Blockchain & Social Media

Blockchain technology is poised to revolutionize social media, offering a decentralized alternative to the current Web2 landscape dominated by giants like Facebook, Twitter, and Instagram. These new platforms leverage blockchain's unique characteristics to empower users with greater control, privacy, and ownership. Enabling a different model for social media with specific features like tokenized economies and decentralized data storage (Wijesekara, 2025). Governance models often involve community participation (Wijesekara, 2025). Challenges include competing with Web2 giants, scalability, and moderation (Wijesekara, 2025).

Examples of Blockchain Social Media Platforms

Diaspora

A non-profit, decentralized social network that runs on free Diaspora software. It emphasizes personal empowerment, freedom, privacy, and decentralization.

Mastodon

A free, open-source social network. It is federated, meaning that it is made up of many different servers (instances) that are independently operated but can communicate with each other.

Minds

A free, open-source social network that focuses on privacy and freedom of speech. It uses blockchain technology to reward users for their contributions.

While blockchain social media platforms are still in their early stages of development, they have the potential to reach a significant user base. Currently, Web2 social media platforms boast billions of users. For example, Facebook has over 2.9 billion monthly active users. In comparison, Mastodon, one of the largest decentralized social networks, has approximately 1.7 million active users. However, the user base of blockchain social media platforms is growing rapidly as more people become aware of the benefits of decentralization and data ownership.

Chapter 2: From Ancient Principles to the Decentralized Web

Having explored the fundamental mechanics and diverse applications of blockchain technology in Chapter 1, we now embark on a journey through its historical landscape. To truly grasp the profound impact and ingenuity of blockchain, it is essential to recognize that this technology did not emerge in a vacuum. Rather, it represents a sophisticated synthesis and evolution of long-standing human practices in record-keeping, security, and reaching consensus, empowered and refined by modern technological advancements, particularly in cryptography and distributed networking. Understanding these historical roots provides crucial context for appreciating how blockchain addresses age-old challenges in novel and powerful ways, paving the path toward a more decentralized future.

The Historical Roots of Blockchain: Ancient Practices, Modern Solutions

The analogy of blockchain as the "New Silk Road" hints at its role in facilitating exchange and collaboration across vast distances, much like the ancient trade routes connected civilizations. However, to truly grasp blockchain's significance, we must look deeper into specific historical practices and understand how blockchain addresses their inherent limitations. By examining how societies throughout history tackled fundamental problems of trust and record integrity, we can see the lineage that led to the decentralized web we are beginning to build today.

Decentralized Record-Keeping: Echoes of Ancient Ledgers

The fundamental principle of maintaining records across multiple, independent parties, rather than relying on a singular, centralized authority, has deep roots in human history. This distributed approach inherently offers greater resilience against data loss, tampering, and corruption – a need that various societies addressed in ingenious ways long before the advent of digital technology.

Redundancy

The Redundancy of Clay Tablets

In ancient Mesopotamia, often considered the cradle of writing and accounting, scribes meticulously recorded commercial transactions, land ownership, and legal agreements on clay tablets. Crucially, these records were often duplicated and stored in different temples, administrative centers, or even within the involved families' possessions (Spar, 2004). This deliberate redundancy served as a safeguard against the accidental destruction of a single record due to fire, flood, or conflict. If one set of tablets was lost or damaged, others could be consulted to verify the information.

Blockchain's Evolution of Redundancy

Blockchain mirrors this principle by distributing the entire ledger across numerous nodes in its network. However, it significantly enhances this resilience through automatic and continuous synchronization. Every addition or modification to the blockchain is instantly replicated across the network, ensuring that the ledger exists simultaneously in thousands (or even millions) of locations globally (DeJeu, 2025). This eliminates the risk of localized data loss affecting the entire record and makes targeted data alteration by a single party practically impossible due to the need to alter the majority of the distributed copies cryptographically.

Accountability

Medieval Guilds: Cross-Referencing for Accountability

Moving forward in history, during the medieval period, craft and merchant guilds played a vital role in regulating trade and ensuring fair practices among their members. To prevent fraud and maintain accountability, guilds often mandated the keeping of separate accounting books by individual members or branches (Mitchell, n.d.). These independent records were then periodically cross-referenced and audited by guild officials or committees. Discrepancies would trigger investigations, ensuring that no single entity could easily manipulate financial information for personal gain.

Blockchain's immutability and transparency build upon this concept of cross-referencing. Once a transaction is recorded on the blockchain, it cannot be altered retroactively (DeJeu, 2025). The distributed nature allows any participant to independently verify the transaction history, effectively acting as a continuous, automated audit. The cryptographic linking of blocks ensures that any attempt to tamper with past records would be immediately evident when compared against the unaltered versions held by other network participants.

Ledgers & Registries

Historical Land Registries: Community-Based Security

Throughout history, systems for recording land ownership have often involved a degree of distribution to enhance security and prevent fraudulent claims. In some societies, records were maintained not just in central government offices but also within local communities, perhaps through publicly accessible registers or even through communal knowledge and witnessed agreements (Park, 2024). This distributed knowledge base made it significantly harder for a single entity to fraudulently alter ownership records without the knowledge or consent of the local community.

Blockchain takes this community-based security to a global, digital scale. The distributed nature of the ledger means that no single entity controls the record of ownership. Instead, the network as a whole validates and maintains the integrity of these records through cryptographic consensus. This eliminates the reliance on potentially fallible or corruptible central authorities and provides a more robust and transparent system for tracking digital assets and, potentially in the future, even physical assets like land titles.

In essence, blockchain is not a radical departure from historical needs and practices. Instead, it represents a powerful evolution, leveraging advancements in cryptography and distributed computing to address the inherent limitations of traditional methods for maintaining secure, transparent, and trustworthy records and facilitating decentralized agreement.

Cryptography

While the concept of blockchain feels distinctly modern, its foundational security mechanism – cryptography – has roots stretching back to ancient civilizations. Blockchain isn't just a technological breakthrough; it's a culmination of centuries of human striving for secure, transparent, and trustworthy systems of information management.

Ancient Origins of Secure Communication

The need for secure communication and record-keeping dates back millennia. Ancient civilizations developed various forms of cryptography, from simple substitution ciphers used in warfare to more complex methods for protecting sensitive information (Schneider, 2024). The scytale used in ancient Sparta and the Caesar cipher are early examples of attempts to ensure confidentiality and integrity. Even the complex Enigma machine used in World War II highlights the ongoing evolution of cryptographic techniques to protect sensitive information.

Blockchain's Leverage of Modern Cryptography

Blockchain takes cryptography to the next level by using sophisticated algorithms like hashing and digital signatures.

Hashing

Hashing algorithms (like SHA-256 used in Bitcoin) are one-way mathematical functions that take an input (any data, like a block of transactions) and produce a fixed-size output (a hash or digest) that is unique to that input (Ahmed, 2023). Even a tiny change in the input data will result in a completely different hash. This creates a unique "fingerprint" of the data, ensuring its integrity. If the data in a block is altered, the hash of that block changes, instantly detectable by the network (Ahmed, 2023).

Digital Signatures

Digital signatures use a pair of cryptographically linked keys – a public key and a private key – to verify the authenticity of information and prevent forgery (Ahmed, 2023). A user signs a transaction with their private key, and anyone can use their corresponding public key to verify that the signature is valid and that the transaction originated from that user. This ensures non-repudiation and trust in the origin of transactions within the network.

Building Trust Through Cryptographic Proof

These cryptographic techniques, while advanced, build upon the fundamental human desire to protect information and ensure its trustworthiness. By making data integrity verifiable and transaction origins attributable (even if pseudonymous), blockchain uses cryptography to establish trust computationally and mathematically, rather than solely relying on trust in intermediaries.

Consensus

The fundamental human need to reach consensus, or achieve agreement within a group, has driven the development of diverse decision-making processes throughout history. Societies have evolved mechanisms ranging from simple majority rule to intricate systems designed to ensure inclusivity and prevent unilateral control. Blockchain's innovative consensus mechanisms represent a digital evolution of these age-old challenges, providing automated ways for a distributed network to agree on a single version of the truth about the state of the ledger.

Tribal Councils Consensus Through Deliberation

In many indigenous cultures, decisions affecting the community were made through extensive deliberation in tribal councils. The emphasis was often on achieving a consensus where all voices were heard and considered, even if it required lengthy discussions and compromises. This approach aimed to ensure the legitimacy and acceptance of decisions within the community, preventing dissent and fostering social cohesion. While blockchain's automated consensus mechanisms differ significantly in their execution, they share the underlying goal of achieving agreement across a diverse group without relying on a central authority.

Quaker Consensus Seeking Unity in Agreement

The Quaker tradition of consensus emphasizes finding unity through open dialogue and seeking solutions that satisfy the conscience of all members. Decisions are not made by majority vote but through a process of collective discernment until a sense of "the way forward" emerges that everyone can support. This highlights the importance of inclusivity and the potential pitfalls of decisions imposed by a dominant group. Blockchain's design, particularly in its consensus mechanisms, aims to create a system where no single entity can unilaterally dictate the state of the ledger. The requirement for a majority (or a specific threshold, depending on the consensus mechanism) of the distributed network to agree on a transaction before it is added ensures that the system reflects a broad consensus rather than the will of a single actor (Park, 2024).

Byzantine Fault Tolerance: Designing for Unreliable Participants

The concept of Byzantine Fault Tolerance (BFT), originating in theoretical computer science, directly addresses the challenge of achieving reliable consensus in a system where some participants may be unreliable, faulty, or even intentionally malicious (Lamport et al., 1982). The "Byzantine Generals Problem" is a classic thought experiment illustrating this challenge: how can several generals surrounding a city coordinate an attack if some of them might be traitors trying to sabotage the plan? Byzantine Fault Tolerance (BFT) algorithms provide solutions to this problem by ensuring that the system can continue to operate correctly even if some nodes fail or act maliciously (Lamport et al., 1982). Blockchain consensus mechanisms, such as Practical Byzantine Fault Tolerance (PBFT) used in some permissioned blockchains, draw inspiration from BFT algorithms to ensure that the network can reach a consensus on the state of the ledger even if some nodes are compromised (Park, 2024).

By examining these historical parallels, we see that blockchain, while a novel technological implementation, builds upon fundamental principles of distributed record-keeping and consensus-building that have been essential for fostering trust and cooperation throughout human history. Its innovation lies in its ability to automate and enhance these principles through cryptography and distributed networking, creating a more resilient, transparent, and potentially more equitable system for managing information and value in the digital age.

The Evolution of the Internet: From Web 1.0 to the Promise and Perils of Web 3.0

Just as blockchain has historical roots in ancient practices, it is also a product of the internet's own evolution. To truly grasp the transformative potential and the inherent complexities of Web 3.0 and its deep connection to blockchain technology, it's essential to understand the internet's journey through its previous iterations.

Web 1.0 (Read-Only Web)

The internet's initial phase, roughly from the early 1990s to the early 2000s, was characterized by static websites, primarily serving as a digital library where users could passively consume information created by a limited number of publishers (Pömer, 2025). Interaction was minimal, often limited to basic hyperlinks navigating between pages. This era, while foundational, offered little in the way of user participation or content creation.

Web 2.0 (Social Web)

The advent of Web 2.0, beginning in the early 2000s, marked a significant shift towards a dynamic and interactive internet. The rise of social media platforms, blogs, wikis, and other user-generated content platforms empowered individuals to become creators and engage in online communities (Pömer, 2025). This era fostered unprecedented connectivity and information sharing. However, it also ushered in an era of increasing centralization. A handful of powerful corporations emerged as gatekeepers, controlling vast amounts of user data, dictating platform terms, and often monetizing user activity in ways that prioritized their own interests. This centralization has led to significant concerns about data privacy, censorship, algorithmic bias, and the concentration of economic power online.

Web 3.0: The Ambition of a Decentralized and Semantic Web

Web 3.0 represents a multifaceted vision for the next evolution of the internet, with decentralization as a core tenet. It aims to move beyond the limitations of Web 2.0 by leveraging technologies like blockchain, semantic web principles (making data more understandable to machines), artificial intelligence, and the Internet of Things (IoT) to create a more user-centric, transparent, and secure online experience (Pömer, 2025). The ambition is to shift power and control away from centralized intermediaries and back into the hands of individuals and communities.

Challenges in Realizing Web 3.0

Portraying Web 3.0 simply as the "decentralized web" oversimplifies the profound changes it seeks to enact and the significant hurdles it faces. True decentralization is not merely a technological shift; it necessitates a fundamental re-architecting of how the internet's infrastructure, applications, and governance operate. This transition involves tackling complex challenges:

Scalability and Performance

As discussed in Chapter 1, many current decentralized technologies, particularly early blockchain iterations, face scalability limitations, struggling to handle the transaction volumes and speeds required for mainstream internet applications (Aldoubaee et al., 2024). Overcoming these performance bottlenecks is crucial for widespread adoption.

Usability and User Experience

Interacting with decentralized applications (dApps) and managing cryptographic keys can be technically challenging for the average user. Creating intuitive and user-friendly interfaces is essential for bridging the gap between the technical underpinnings of Web 3.0 and mass adoption (Hahn, 2025).

Regulation and Governance

The decentralized nature of Web 3.0 poses significant challenges for existing regulatory frameworks. Establishing clear and effective governance mechanisms that can address issues like illegal content, fraud, and consumer protection without undermining the principles of decentralization is a complex and ongoing debate (Silverbreit, 2025).

Interoperability

Seamless interaction between different decentralized platforms and protocols is vital for a cohesive Web 3.0 ecosystem. Achieving interoperability standards and facilitating the exchange of data and value across different blockchains and dApps remains a significant technical and strategic challenge (Kotey et al., 2024).

Security and Smart Contract Vulnerabilities

While blockchain itself is inherently secure, the smart contracts that power many decentralized applications are susceptible to vulnerabilities and exploits. Ensuring the security and reliability of these contracts is paramount for building trust in the Web 3.0 ecosystem (Huang et al., 2023).

The Transition from Web 2.0

Moving from the established infrastructure and user habits of Web 2.0 to a decentralized paradigm requires significant incentives and a compelling value proposition for both users and developers. Overcoming the powerful network effects and ingrained behaviors of centralized platforms will be a gradual and potentially disruptive process (Pömer, 2025).

Fulfilling the Internet's Original Vision: Reclaiming Decentralization and Empowerment

The initial promise of the internet was indeed a decentralized frontier, a space for open communication and individual empowerment, free from the control of centralized entities. The rise of Web 2.0, while bringing immense benefits in terms of connectivity and user participation, inadvertently led to the consolidation of power and data within a few dominant corporations, often at the expense of user privacy and autonomy.

Blockchain technology offers a potential pathway to reignite this original vision by providing the foundational infrastructure for a more decentralized internet. Its inherent characteristics directly address some of the key limitations of Web 2.0:

Data Ownership and Control

Blockchain-based systems empower individuals with greater control over their data. Instead of their information being siloed and controlled by centralized platforms, users can own their data, decide how and with whom it is shared, and potentially even monetize it directly (Mukherjee et al., 2025). This shift in data ownership is a fundamental departure from the data extraction model prevalent in Web 2.0.

Open and Permissionless Platforms

Blockchain facilitates the development of open and permissionless platforms, reducing reliance on centralized intermediaries that can act as gatekeepers, censor content, or impose arbitrary rules (DeJeu, 2025). Decentralized applications built on blockchain can operate autonomously, governed by smart contract logic and community consensus, fostering a more level playing field for innovation and participation.

Community-Driven Governance and Value Creation

Blockchain enables the creation of decentralized autonomous organizations (DAOs) and tokenized ecosystems that empower communities to collectively govern platforms and share in the value they create (Lamport et al., 1982). This fosters a more inclusive and democratic internet where users are not just consumers but also stakeholders and active participants in shaping the online landscape.

What Does it Mean to Own the Internet?

One key difference with Web 3.0 compared to Web 1 & 2 is that any individual can own a piece of the internet. You may ask, what does this mean? Blockchain technology (Web 3) is emancipatory. In Web 3.0, individuals can become owners of the web in a few key ways:

1. **Decentralized Identity:** Individuals can create and control their own digital identities on a blockchain. This allows people to own and manage their

personal data, rather than having it controlled by centralized platforms.

2. **Tokenized Ownership:** Some Web3 platforms use tokens to represent ownership of digital assets or communities. By holding tokens, individuals can directly own a piece of the platform and participate in its governance.

3. **Decentralized Governance:** Web3 often involves DAOs that allow token holders to participate in decision-making. This gives individuals a direct voice in how the platform or community is run.

4. **Creation and Ownership of Digital Assets:** Individuals can create and own unique digital assets in the form of NFTs (non-fungible tokens). These NFTs can represent ownership of digital art, collectibles, or other forms of digital content.

5. **Participation in the Development and Maintenance of the Web:** Web3 encourages open-source development and community involvement in building and maintaining the web. This allows individuals to contribute to the growth and evolution of the web, rather than being passive consumers.

Web3 aims to shift ownership and control of the web from centralized platforms to individual users (Vendette & Thundiyil, 2023). This is achieved through a combination of blockchain technology, tokenization, and decentralized governance models. By empowering individuals to own their data, participate in decision-making, and contribute to the development of the web, Web3 aims to create a more democratic and user-centric internet.

While the vision of a fully realized Web 3.0 is still in its nascent stages and faces significant technical, social, and regulatory hurdles, blockchain technology provides a crucial building block for achieving a more decentralized, transparent, and user-empowered internet – one that more closely aligns with the original ideals of its creators. Understanding the complexities and challenges involved in this transition is just as important as recognizing its immense potential.

The Genesis of Bitcoin: A Response to Crisis and a Vision for the Future

The direct precursor to modern blockchain technology as we know it was the creation of Bitcoin. Understanding its origins provides essential context for the development of the entire blockchain ecosystem and the motivations behind building a decentralized digital currency.

The 2008 Financial Crisis: A Catalyst for Change

The 2008 financial crisis exposed the fragility of the traditional financial system and severely eroded public trust in centralized financial institutions. The reckless practices of banks and the subsequent government bailouts fueled a desire for alternative financial systems that were more transparent, accountable, and resilient, operating outside the control of traditional authorities (Weinberg, 2013).

In this climate of distrust, a pseudonymous individual (or group) known as Satoshi Nakamoto introduced Bitcoin in a white paper titled *Bitcoin: A peer-to-peer electronic cash system* in October 2008 (Nakamoto, 2008). Nakamoto proposed a decentralized digital currency that would operate independently of central banks and governments, offering a potential solution to the problems plaguing the traditional financial system by enabling peer-to-peer electronic transactions without relying on a trusted third party.

Key Innovations of the Bitcoin Whitepaper

The Bitcoin whitepaper outlined several groundbreaking innovations that together formed the first functional blockchain:

Decentralized Ledger

It introduced the concept of a decentralized, public ledger (the blockchain) that would record all transactions in a transparent and tamper-proof manner, eliminating the need for a central authority (ImmuneBytes, 2024).

Proof-of-Work Consensus

To secure the network and prevent fraud (specifically, the double-spending problem), Bitcoin employed a Proof-of-Work (PoW) consensus mechanism, requiring miners to expend computational power to add new blocks to the blockchain (Conway, 2022; Nakamoto, 2008).

Cryptographic Security

Bitcoin utilized cryptographic techniques, such as hashing and digital signatures, to ensure the integrity and authenticity of transactions (Ahmed, 2023; Nakamoto, 2008).

Limited Supply

Bitcoin introduced the concept of a limited supply of coins (capped at 21 million), creating digital scarcity and potentially protecting against inflation (Nakamoto, 2008).

Motivations Behind Bitcoin

The motivations behind Bitcoin's creation were rooted in the desire for a more robust and trustworthy financial system:

Decentralization

Creating a financial system free from central control and censorship was a key motivation (Nakamoto, 2008).

Transparency and Accountability

Bitcoin's public ledger promoted transparency and accountability in financial transactions (Nakamoto, 2008).

Security and Immutability

The use of cryptography and PoW ensured the security and immutability of the blockchain, protecting against fraud and manipulation (Conway, 2022; Nakamoto, 2008).

Financial Inclusion

Bitcoin aimed to provide access to financial services for those who were excluded from the traditional banking system (Nakamoto, 2008).

Bitcoin's creation marked a pivotal moment in the history of finance and technology. It not only introduced the first successful cryptocurrency but also laid the foundation for the development of blockchain technology as a whole. Bitcoin's core principles of decentralization, transparency, and security have inspired countless other blockchain projects and applications across various industries.

The birth of Bitcoin was a direct response to the 2008 financial crisis and a powerful testament to the desire for a more decentralized and transparent financial system. By analyzing the Bitcoin whitepaper and understanding its key innovations, we gain a deeper appreciation for the origins of blockchain technology and its potential to reshape our society.

Beyond Bitcoin: Charting the Evolution of Blockchain Technology

The emergence of Bitcoin served as a powerful catalyst, transforming the perception of blockchain technology from a niche concept to a mainstream phenomenon. IT professionals, entrepreneurs, investors, and business leaders, initially intrigued by

Bitcoin's decentralized nature, began to recognize the broader potential of its underlying blockchain technology. This curiosity fueled a surge in experimentation and innovation, leading to the discovery of blockchain's applicability in areas far beyond digital currencies.

The Rise of Ethereum and Smart Contracts

A major leap in blockchain's evolution came with the introduction of Ethereum. In 2015, Vitalik Buterin introduced Ethereum, a blockchain platform that went beyond Bitcoin's focus on cryptocurrency (Buterin, 2014). Buterin envisioned a more general-purpose blockchain that could support a wider range of applications.

Ethereum's key innovation was the introduction of smart contracts – self-executing contracts with the terms of the agreement directly written into code (Buterin, 2014). These contracts automatically execute when predefined conditions are met, removing the need for intermediaries.

Smart contracts enabled the development of decentralized applications (dApps), which operate on a blockchain network without central control. This opened up a vast array of possibilities for blockchain technology, extending its reach far beyond finance into areas like decentralized finance (DeFi), supply chain management, digital identity, and decentralized marketplaces (Buterin, 2014).

While Bitcoin's Proof-of-Work (PoW) consensus mechanism was groundbreaking, its limitations in terms of scalability and energy consumption became apparent as the network grew (Conway, 2022; Omar et al., 2025). To address these limitations, Proof-of-Stake (PoS) emerged as a more energy-efficient and potentially scalable alternative (Park, 2024). In PoS, validators stake their own cryptocurrency to participate in the consensus process, replacing the energy-intensive mining of PoW.

The blockchain ecosystem continues to explore and develop new consensus mechanisms, such as Delegated Proof-of-Stake (DPoS), Proof-of-Authority (PoA), and Practical Byzantine Fault Tolerance (PBFT), each with its own trade-offs and advantages in terms of decentralization, speed, and security (GeeksforGeeks, 2025; Park, 2024; Roy, 2023). This ongoing evolution highlights the continuous efforts to improve blockchain's efficiency, security, and scalability.

Building Blocks of Blockchain: Exploring the Technological Precursors

To fully appreciate the technical elegance of blockchain, it's helpful to understand some specific technological concepts that directly influenced its development.

Hashcash: A Proof-of-Work Pioneer

Developed by Adam Back in 1997, Hashcash aimed to deter email spam by requiring senders to perform a small amount of computational work before sending an email (Back, 2002). This "proof-of-work" system made sending spam more costly and time-consuming, discouraging bulk spamming. Bitcoin adopted a similar proof-of-work system to secure its network and prevent malicious attacks, requiring miners to expend computational power to solve complex mathematical problems to add new blocks to the blockchain (Nakamoto, 2008). This connection to Hashcash demonstrates how early innovations in cryptography and distributed systems directly influenced blockchain's design.

Merkle Trees: Efficient Data Verification

Invented by Ralph Merkle in 1979, Merkle trees are a tree-like data structure that allows for efficient verification of large amounts of data (Merkle, 1980). Each "leaf" of the tree represents a piece of data (like a transaction), and each "node" above it represents a hash of its child nodes. The "Merkle root" at the top is the hash of the entire set of data. Blockchain utilizes Merkle trees to summarize and verify the integrity of transactions within a block. This allows for efficient verification of data without needing to examine every single transaction, improving efficiency and scalability.

This concept from distributed computing explores the challenge of reaching consensus in a system where some participants may be unreliable or malicious (Lamport et al., 1982). The "Byzantine Generals Problem" is a classic thought experiment illustrating this challenge: how can several generals surrounding a city coordinate an attack if some of them might be traitors trying to sabotage the plan? Byzantine Fault Tolerance (BFT) algorithms provide solutions to this problem by ensuring that the system can continue to operate correctly even if some nodes fail or act maliciously (Lamport et al., 1982). Blockchain consensus mechanisms, such as Practical Byzantine Fault Tolerance (PBFT) used in some permissioned blockchains, draw inspiration from BFT algorithms to ensure that the network can reach a consensus on the state of the ledger even if some nodes are compromised (Park, 2024).

These technological precursors highlight the rich history of innovation that led to the development of blockchain. Hashcash's proof-of-work system, Merkle trees' efficient data verification, and BFT algorithms' consensus-building capabilities all contributed to the foundation upon which blockchain technology was built. Recognizing these building blocks allows us to better understand blockchain's potential to revolutionize various industries and create a more secure, transparent, and decentralized internet.

Growth of the Blockchain Ecosystem

Following the breakthroughs of Bitcoin and Ethereum, the blockchain ecosystem has expanded significantly, driven by ongoing innovation and increasing interest across various sectors. The ecosystem now includes numerous blockchain platforms beyond Bitcoin and Ethereum, each with their own focus and characteristics, designed for different use cases and prioritizing different trade-offs (Gillis, 2023; SNS Insider, 2025). Examples include enterprise-grade platforms like Hyperledger Fabric (permissioned networks), R3 Corda (financial institutions), and Quorum (enterprise Ethereum).

A wide range of tools and technologies have been developed to support blockchain development, deployment, and adoption. These include digital wallets for securely storing cryptocurrencies and digital assets, exchanges for trading, and development frameworks and libraries for building blockchain applications (SNS Insider, 2025). The blockchain ecosystem is also supported by a vibrant global community of developers, entrepreneurs, researchers, and enthusiasts. Online forums, conferences, hackathons, and open-source projects foster collaboration and drive continuous innovation (SNS Insider, 2025).

Recognizing Key Contributors to Blockchain

While blockchain technology is often perceived as a purely technical phenomenon, it's important to acknowledge the contributions of key figures and organizations that have played a crucial role in shaping the blockchain landscape.

Vitalik Buterin

The Visionary Behind Ethereum: Vitalik Buterin, a Russian-Canadian programmer, became interested in Bitcoin in its early days but recognized its limitations (Buterin, 2014). In 2013, he published the Ethereum whitepaper, proposing a new platform with a general-purpose scripting language to support a wider range of applications through smart contracts (Buterin, 2014). Buterin remains a leading figure, advocating for decentralized technologies.

Gavin Wood Architecting the Future of Blockchain

Gavin Wood, a British computer scientist, was a co-founder of Ethereum and developed the Solidity programming language for smart contracts (Buterin, 2014). He later founded Polkadot, a multi-chain network focused on interoperability, and the Web3 Foundation, supporting decentralized technologies (Kotey et al., 2024). Wood's work addresses key challenges in blockchain scalability and interoperability.

The Ethereum Foundation Supporting a Decentralized Ecosystem

The Ethereum Foundation is a non-profit organization dedicated to fostering the growth and evolution of the Ethereum platform, supporting its development, research, and community building. Its origins are deeply intertwined with the Ethereum founding team (Buterin, 2014).

The Bitcoin Foundation Advocating for Bitcoin's Growth

The Bitcoin Foundation, founded in 2012 by key figures like Peter Vessenes and Gavin Andresen, is a non-profit organization promoting the use and development of Bitcoin, playing a key role in its early standardization and adoption (Wikipedia, n.d.).

Highlighting the contributions of key figures and organizations in the blockchain space reminds us that blockchain is not just about code and algorithms but also about the people who are driving its development and shaping its future. By recognizing these individuals and organizations, we can better appreciate the collaborative effort and vision that are driving the blockchain revolution.

The evolution of blockchain technology has been remarkable, expanding from its origins in cryptocurrency to a wide range of applications across various industries. The emergence of smart contracts, the development of diverse consensus mechanisms, and the growth of the blockchain ecosystem demonstrate the continuous innovation and adaptation driving this technology forward.

This chapter has offered a foundational overview of blockchain's origins and its potential to revolutionize how we interact, transact, and govern ourselves in the digital age. As we've seen, this complex and evolving system is built upon a rich history of human practices and technological innovation. Its ultimate success and impact, however, hinge on humans. Unlike traditional technologies with pre-defined user roles, blockchain thrives on active participation and user trust. The true power of blockchain lies not just in its technical capabilities but also in its ability to empower individuals, foster collaboration, and promote transparency and trust. By understanding its potential and embracing its possibilities, we can collectively shape a future where technology serves humanity and creates a more equitable and decentralized internet. To further explore this, the next chapter will bridge the gap between these technical foundations and the human factors that determine how blockchain is adopted, used, and perceived. It will delve into behavioral science to understand how our cognitive biases, motivations, and decision-making processes interact with blockchain technology, shaping its trajectory and our experience of the decentralized web, acknowledging the historical roots that paved the way for this transformative technology.

Chapter 3: The Human Side of Blockchain

Blockchain technology, often discussed in terms of its technical architecture, cryptographic security, and economic implications, is fundamentally a human-driven system. Its creation was a response to human needs and distrust, its adoption is shaped by human motivations and behaviors, and its future impact will be determined by human choices and interactions. This chapter argues that truly unlocking blockchain's transformative potential requires looking beyond the code to understand the complex interplay between this technology and the human mind (Vendette & Thundiyil, 2023). By integrating insights from behavioral science, cognitive psychology, and social science – fields illuminated by pioneers like Daniel Kahneman, Amos Tversky, and further enriched by perspectives from researchers exploring human decision-making, social dynamics, and the impact of technology – we can develop a deeper understanding of the "human algorithm" that underlies the decentralized web. This understanding is crucial for designing blockchain systems that are not only technically robust but also intuitive, trustworthy, equitable, and genuinely aligned with human needs and aspirations, ultimately contributing to the vision of a more equitable and sustainable future explored in Chapter 4.

Behavioral science, the study of human behavior and decision-making, provides the missing link in unlocking blockchain's full potential. This chapter delves into how insights from behavioral science can inform the design and implementation of blockchain solutions. By understanding how people think and act, we can create blockchain systems that are not only technically sound but also aligned with human needs and motivations. While blockchain is a groundbreaking technology, behavioral science teaches us that flawed judgment is a normal part of human decision-making. While traditional economics assumes rationality, behavioral economics incorporates these flaws—biases and mental shortcuts—to foster awareness and better choices. Applying behavioral insights when designing blockchain products, governance, and communities can help ensure users make clearer decisions that are fair, ethical, and truly aligned with their own interests and priorities. Humans are not entirely rational creatures. We are susceptible to cognitive biases that can cloud our judgment and influence our perception of new technologies. Biases are cognitive shortcuts or mental tendencies that can lead to systematic errors in judgment and decision-making. They often arise from our attempts to simplify complex situations or rely on past experiences, but they can result in irrational choices.

Traditional economic models often assume humans are purely rational actors, making decisions based on perfect information to maximize their utility. However, behavioral economics, pioneered by Kahneman and Tversky, demonstrates that human judgment is prone to systematic "flaws" – cognitive biases, heuristics (mental shortcuts), and emotional influences – that lead to predictable irrationality (Kahneman & Tversky, 1974; Kahneman & Tversky, 1979). As Kahneman elaborated throughout his work, including *Thinking, Fast and Slow*, our minds operate with two systems: the rapid, intuitive, and emotional "System 1," and the slower, deliberate, and logical "System 2" (Kahneman, 2011). Navigating the complex, novel landscape of blockchain technology often overwhelms System 2, leading users to default to error-prone System 1 thinking, making them particularly susceptible to these biases.

Simultaneously, as Yuval Noah Harari reminds us in *Sapiens*, our capacity for large-scale human cooperation is built upon shared "imagined orders" – collective beliefs and narratives that, while not objectively physical realities, structure our societies, institutions, and even our perception of value (Harari, 2014). Blockchain technology, with its reliance on network consensus and tokenized value, is a powerful new form of imagined order. Understanding how our inherent cognitive biases (Kahneman, Tversky) interact with the formation and maintenance of these digital imagined orders (Harari) is fundamental to building responsible and effective blockchain ecosystems that can gain widespread human trust and adoption.

Understanding Our Biases in the Blockchain Era: The Cognitive Landscape of Web3

To truly understand how individuals perceive, interact with, and make decisions within the novel digital landscape of blockchain, we must apply the psychological lenses provided by behavioral economics and cognitive psychology. These biases are not random errors; they are systematic patterns of thought that arise from our brains' attempts to simplify complexity or rely on easily accessible information (Kahneman, 2011, Kahneman & Tversky, 1974). In the context of blockchain, where complexity is high, information asymmetry can still exist despite transparency, and the stakes can be significant, these biases have profound consequences for adoption, security, governance, and the perception of value.

Let's delve into some key biases and explore their manifestations in blockchain user behavior, drawing on research in behavioral economics and human-computer interaction in novel digital environments.

Prospect Theory and Loss Aversion: Navigating Volatility and Digital Wealth

Prospect theory describes how individuals evaluate potential gains and losses relative to a reference point, with losses having a significantly greater psychological impact than equivalent gains (loss aversion) (Kahneman & Tversky, 1979). This makes people risk-averse regarding potential gains but risk-seeking to avoid certain losses. Research suggests this asymmetry in how we value gains and losses is a fundamental aspect of human decision-making under uncertainty (Thaler, 2015). Harari might link this heightened sensitivity to loss to our evolutionary history, where avoiding loss (of resources, status, life) was critical for survival in uncertain environments (Harari, 2014).

Manifestation in Blockchain Contexts

The volatile nature of cryptocurrency markets and the high-stakes interactions in decentralized finance (DeFi) make the Web3 space fertile ground for loss aversion.

In volatile markets, loss aversion strongly contributes to the "HODL" (Hold On for Dear Life) phenomenon, particularly when asset values drop below an individual's purchase price (their reference point). The pain of realizing a loss can cause users to irrationally hold onto depreciating assets rather than cutting losses, even when System 2 analysis would suggest reallocating capital (Kanga, 2023).

Conversely, sudden market drops can trigger System 1 fear and loss aversion, leading to impulsive panic selling, locking in losses to avoid the *potential* for even larger future losses, even if the rational (System 2) analysis doesn't support it. Users providing liquidity in DeFi pools face "impermanent loss," a potential loss relative to simply holding the assets. The abstract and often complex nature of calculating impermanent loss makes it difficult for System 1 to grasp, while the *potential* for loss triggers System 1 avoidance, creating a psychological barrier to DeFi participation despite potential gains (Schär, 2021). In another scenario, contributors to DAOs might be hesitant to propose or fund initiatives perceived as risky due to loss aversion – not just financial loss, but the loss of reputation or social standing within the community if the project fails. This can stifle innovation and risk-taking necessary for growth (Hsieh, 2017).

Behavioral Science-Informed Design & Interventions

Understanding loss aversion is crucial for designing user interfaces and communication strategies that promote more rational decision-making in volatile or high-risk blockchain environments. Platforms can use behavioral insights to frame potential outcomes more effectively. Instead of just displaying current loss, show potential future scenarios or use language that emphasizes percentage change rather than absolute value (framing effect) (Kahneman, 2011). Clearly communicate the *mechanisms* that protect against loss (e.g., liquidation thresholds in lending protocols).

Offering optional automated tools (e.g., stop-loss orders, rebalancing features in DeFi) can help users make more rational decisions by pre-committing them to actions based on System 2 planning, bypassing System 1's emotional response during volatility. Educational content and platform messaging can frame blockchain engagement around long-term potential and utility rather than short-term price fluctuations, anchoring user perspective away from daily volatility that triggers loss aversion (Kanga, 2023).

Availability Bias: The Shadow of Scams and the Glare of Hype

The availability heuristic leads us to overestimate the likelihood of events that are easily recalled or vivid in our memory (Kahneman & Tversky, 1974). Sensational, recent, or emotionally charged events (like hacks, scams, or parabolic price pumps) are highly "available." Research shows that vivid anecdotes often outweigh statistical data in influencing risk perception (Tversky & Kahneman, 1973). Harari highlights how powerful narratives, especially those involving danger or rapid success, become embedded in our shared consciousness, shaping our perceived reality and influencing collective behavior (Harari, 2014).

Manifestation in Blockchain Contexts

The blockchain and cryptocurrency space has, unfortunately, been associated with its share of scams, hacks, and rug pulls, alongside stories of rapid wealth creation. These often-sensationalized events can become highly "available" in the minds of both newcomers and seasoned participants, leading to a skewed perception of the overall risk and ethical landscape.

Frequent news reports and social media discussions about cryptocurrency scams, rug pulls, and exchange hacks make these negative events highly available.

This leads many, particularly newcomers, to overestimate the overall risk of the blockchain space, perceiving it as inherently dangerous and untrustworthy, regardless of the statistical reality of the vast number of legitimate projects and transactions (Gorkhali & Chowdhury, 2022). This "availability cascade" can create unwarranted fear and stifle rational adoption.

Similarly, highly publicized stories of individuals making fortunes from specific coins or NFTs become extremely available. This fuels FOMO, leading users to engage in impulsive, high-risk investments based on the readily available narrative of easy wealth, without conducting necessary System 2 due diligence, overlooking more probable negative outcomes (Lango, 2021).

The readily available narratives of quick gains or project hype on platforms like X or Discord can prevent users from engaging in the effortful System 2 task of researching project fundamentals, security audits, or team credibility. They rely on the easy System 1 acceptance of the compelling, available story.

Behavioral Science-Informed Design & Interventions

Countering availability bias requires proactive information management and design choices that encourage more deliberate processing. Platforms can counter availability bias by curating and presenting balanced information, highlighting successful use cases, security features, and educational resources alongside necessary risk disclosures (Gorkhali & Chowdhury, 2022). Making reliable, on-chain data easily accessible and understandable can provide users with a more accurate picture than sensationalized anecdotes. Clear data dashboards can help users engage System 2 for more informed risk assessment (Xu & Livshits, 2018). Implementing clear warning signals or "speed bumps" before high-risk actions (like investing in unaudited smart contracts or highly volatile assets) can nudge users to pause and engage System 2, mitigating impulsive decisions driven by available hype or fear (Kahneman, 2011).

Excessive Coherence and Confirmation Bias: Echo Chambers and Unchallenged Narratives in Communities

Also known as the narrative fallacy, this is our tendency to seek, interpret, and remember information that confirms our pre-existing beliefs, and to construct coherent narratives, even if it means ignoring contradictory evidence (Kahneman, 2011). We dislike cognitive dissonance and strive for internal consistency. Research shows that confirmation bias is particularly strong in emotionally charged or identity-linked contexts (Nickerson, 1998). Harari points out that this need for coherent narratives underlies our shared "imagined orders," enabling group cohesion but also making us resistant to information that threatens these fictions (Harari, 2014).

Manifestation in Blockchain Contexts

In the often highly passionate and ideologically driven communities within the Web3 space, excessive coherence and confirmation bias can create echo chambers where dissenting opinions are suppressed, critical questioning is discouraged, and the actions of influential figures or the prevailing group narrative go unchallenged, hindering accountability and balanced decision-making. Blockchain communities, often formed around specific projects or ideologies, are highly susceptible to confirmation bias. Users primarily consume information that confirms their belief in their chosen project's success or the flaws of competing projects. This creates echo chambers on platforms like Discord, Telegram, or X, where dissenting opinions are dismissed, and critical questioning is viewed as an attack on the shared "imagined order" of the community (Nickerson, 1998).

In decentralized governance, excessive coherence can lead to groupthink. If a dominant faction or influential token holder promotes a proposal, confirmation bias can lead other voters to seek information supporting it and ignore valid criticisms, potentially resulting in flawed decisions or the entrenchment of power structures, despite the promise of decentralization (Hsieh, 2017). Users deeply invested in a project (financially or emotionally) may selectively ignore or rationalize away warning signs – unaudited code, missed roadmap deadlines, controversial team behavior – because this information conflicts with their confirmed belief in the project's success narrative (Kahneman, 2011).

Behavioral Science-Informed Design & Interventions

Counteracting excessive coherence and confirmation bias requires designing platforms and community structures that actively encourage exposure to diverse viewpoints and critical thinking. Designing DAO governance platforms with features that explicitly require presentation of opposing viewpoints or allocate space for structured debate before voting can encourage exposure to contradictory information (Hsieh, 2017). Implementing secure, anonymous reporting or suggestion channels within platforms or communities can allow users to raise concerns or share dissenting opinions without fear of social repercussions from the dominant narrative (Reuter et al., 2020). Platform design can encourage users to access information from a variety of sources, rather than solely relying on internal community channels. Features that highlight external audits or independent analyses can challenge internal coherence.

Other Relevant Biases

Beyond these core three, numerous other cognitive biases influence blockchain interactions. For example, the Endowment Effect (valuing something more simply because we own it) can make users irrationally attached to their crypto assets or NFTs (Kahneman et al., 1991). Anchoring Bias can lead users to fixate on initial prices or valuations when assessing asset worth (Ahmed, 2023). Mental Accounting can cause users to treat different crypto holdings differently based on how they were acquired (e.g., treating airdropped tokens as "free money" and taking higher risks) (Thaler, 1985). Understanding these biases provides a more complete picture of human behavior in the blockchain space.

The Theory of Blockchain Psychology: A Framework for Human-Centered Design

Building on the understanding of these fundamental biases and cognitive tendencies, we can introduce the concept of the Theory of Blockchain Psychology. This theory posits that the successful integration and widespread adoption of blockchain technology depend fundamentally on its alignment with human behavior, cognition, and social dynamics. It moves beyond a purely technical view, recognizing that blockchain's true potential lies in its capacity to resonate with and enhance human experiences by building upon our inherent social and cognitive architectures, while also mitigating the pitfalls of our predictable irrationality (Kahneman, Tversky) and the challenges of managing shared beliefs (Harari).

Underlying Assumptions

Blockchain Adoption is Not Purely Rational

Decisions to adopt, use, or invest in blockchain technologies are not solely based on objective technical or economic factors but are significantly influenced by cognitive biases, emotions, and social factors.

Human Factors Are Predictable and Designable

While individual behavior can be complex, the systematic patterns of human cognition and social interaction are predictable to a significant degree (as demonstrated by behavioral science) and can be influenced through careful system design and communication.

Technology and Human Behavior are Interdependent

Blockchain technology is not a neutral tool; its design influences human behavior, and human behavior, in turn, shapes the development, adoption, and impact of blockchain.

Shared Beliefs Shape Digital Realities

The success of decentralized systems relies on the collective belief of participants in the system's rules, security, and value (Harari's imagined orders), and these beliefs are susceptible to behavioral influences.

Core Components of Blockchain Psychology

The theory identifies several core components crucial for understanding the human-blockchain intersection:

Cognitive Alignment

Blockchain solutions must be designed with a deep understanding of human cognitive processes. This involves minimizing cognitive load (Kahneman's System 2 effort) through intuitive UI/UX, clear mental models of complex processes (like transactions, wallets, smart contracts), and simplifying technical jargon (Kahneman, 2011). As Harari highlights, creating comprehensible narratives around blockchain's function is crucial for building a shared understanding and trust – a process heavily influenced by our System 1 interpretation and susceptibility to biases (Harari, 2014). Designing for cognitive alignment helps users make more informed decisions and reduces errors.

Emotional Resonance

Trust, engagement, and adoption are deeply emotional. Blockchain design should evoke positive emotions (trust through transparency/security, empowerment through control, belonging in communities) while mitigating negative ones (fear of loss/complexity, anxiety from volatility) (Thaler, 2015). Understanding emotional heuristics (Kahneman) and our innate social needs (Harari's emphasis on cooperation and belonging) is key (Harari, 2014; Kahneman, 2011). Designing for emotional resonance fosters user comfort, confidence, and loyalty.

Social Dynamics and Governance

Blockchain-based communities and DAOs are complex social systems. Applying principles of social psychology – group behavior, cooperation theory, conflict resolution, norm formation – is vital for designing effective governance (Hsieh, 2017). Harari's concept of "imagined orders" provides the macro context for understanding how the rules and norms of these digital societies gain legitimacy and influence behavior, a process where individual biases like confirmation bias and groupthink play a significant role (Harari, 2014; Nickerson, 1998). Designing for positive social dynamics promotes collaboration, reduces conflict, and strengthens community cohesion.

Cultural Context and Adoption

Adoption patterns are shaped by cultural values, norms, and existing social structures. Blockchain psychology recognizes that behavioral responses and the influence of specific biases can vary across cultural contexts (Hofstede, 2001). Successful solutions must be culturally sensitive and integrated into existing social landscapes (Harari, 2014). Designing for cultural context ensures relevance, acceptance, and equitable access across diverse populations.

Ethical Considerations and Human Values

Blockchain design must proactively consider ethical implications, aligning with fundamental human values like fairness, privacy, accessibility, and sustainability. This involves understanding how biases can lead to unethical outcomes (e.g., algorithmic bias perpetuating discrimination) and designing systems to mitigate these risks, prioritizing human well-being within the digital imagined order (Floridi, 2013). Designing for ethical considerations ensures that technology serves humanity's best interests.

A Model for Understanding Blockchain Adoption and Use

A potential model for understanding the interplay between psychological factors, blockchain characteristics, and user outcomes could be represented as follows:

Psychological Factors

- Cognitive Processes (Perception, Decision-Making, Mental Models)
- Emotional Responses (Trust, Fear, Empowerment, Belonging)
- Social Dynamics (Group Behavior, Cooperation, Conflict)
- Cultural Context (Values, Norms, Beliefs)

↓ ↑ **(Influence each other)**

Blockchain Characteristics

- Decentralization (Control, Transparency, Security)
- Immutability (Trust, Auditability)
- Tokenization (Incentives, Ownership)
- Smart Contracts (Automation, Trustless Execution)
- Community Governance (Participation, Equity)

↓**(Both directly impact user outcomes)**

User Outcomes

- Adoption Rate
- User Engagement and Satisfaction
- Trust and Security Perceptions
- Community Participation and Cohesion
- Positive Social Impact (e.g., Financial Inclusion, Transparency)

Predictive Power

The Theory of Blockchain Psychology suggests that by analyzing a blockchain system or application through the lens of these core components and the relevant behavioral principles (biases, heuristics, social dynamics), we can gain predictive power regarding:

User Adoption Rates: Identifying potential cognitive or emotional barriers in the user journey can help predict where users will drop off or hesitate.

Vulnerability to Manipulation: Understanding how biases like availability or confirmation bias manifest in a specific platform's design or community structure can highlight vulnerabilities to scams, misinformation, or undue influence.

Effectiveness of Governance Mechanisms: Analyzing governance structures through the lens of social dynamics and cognitive load can predict participation rates, the likelihood of groupthink, or the potential for power imbalances.

Success of Incentive Structures: Applying principles from Prospect Theory and motivation research can help predict how users will respond to tokenized rewards or penalties.

Points of Friction and Error: Identifying areas of high cognitive load or complex decision points can predict where users are likely to make mistakes or abandon a process.

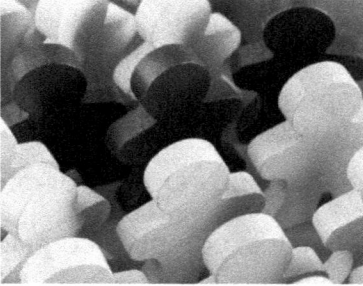

Core Components of The Theory
of Blockchain Psychology

• Cognitive Alignment
• Emotional Resonance
• Social Dynamics and Governance
• Cultural Context and Adoption
• Ethical Considerations and Human Values

Connecting Biases to System Noise

Building on the concept of individual biases, we can understand how these micro-level cognitive tendencies contribute to macro-level phenomena within blockchain ecosystems, specifically "system noise." As described by Daniel Kahneman, Oliver Sibony, and Cass R. Sunstein in *Noise: A flaw in human judgement*, noise refers to unwanted variability and inconsistency in judgments that should ideally be identical (Kahneman et al., 2021). In complex, interactive systems like blockchain networks and their associated communities, individual biases aggregate and interact with the system's architecture to create collective "system noise" – the inherent variability and inconsistencies in user behavior, interactions, and the spread of information that emerge from the platform's features and the collective dynamics of its users, distorting clear signals and making outcomes less predictable.

Prospect Theory & Emotional Noise

Loss aversion fuels emotional noise. The fear associated with potential financial losses (or the pain of realized losses) creates volatility in user sentiment and decision-making, leading to unpredictable cascades of buying or selling or irrational adherence to failing projects, adding "static" to market behavior and community dynamics (Kahneman et al., 2021; Kanga, 2023).

Availability Bias & Cognitive Noise

Availability bias contributes to cognitive noise by flooding the system with easily recalled, often sensationalized, information (scams, pumps). This distracts from relevant signals, distorts overall risk perception, and makes it harder for users to make consistently rational decisions based on underlying fundamentals, introducing variability in judgments and contributing to market volatility or irrational investment trends (Gorkhali & Chowdhury, 2022; Kahneman et al., 2021; Lango, 2021).

Excessive Coherence & Social Noise

Excessive coherence generates social noise in the form of echo chambers and unchallenged group narratives. This creates pockets of highly correlated, potentially inaccurate beliefs within the system, stifling diverse viewpoints and leading to inconsistent responses to information or events across the broader ecosystem (Kahneman et al., 2021; Nickerson, 1998).

Behavioral Intervention for System Noise Solutions

Mitigating system noise requires designing blockchain platforms and communities with an awareness of how individual biases aggregate. Solutions are not just technical, but behavioral, focusing on improving "decision hygiene" within the system (Kahneman et al., 2021):

Information Architecture & Framing

Design platforms to curate information, present data clearly (reducing cognitive load), and frame choices in ways that encourage more deliberate (System 2) processing, countering the impact of availability bias and emotional noise (Gorkhali & Chowdhury, 2022; Kahneman, 2011).

Structured Interaction Mechanisms

Implement features that introduce friction or structure into decision-making processes (e.g., mandatory cooling-off periods for large trades, structured proposal templates in DAOs, requiring citation of sources in community discussions) to counteract impulsive System 1 responses and excessive coherence (Hsieh, 2017; Kahneman et al., 2021).

Diversity & Inclusion

Design governance mechanisms and community platforms that actively promote diverse participation and viewpoints, using mechanisms like quadratic voting (which reduces the power of large holders) or incentivizing contributions from underrepresented groups, countering the effects of excessive coherence and potential biases in power structures (Hsieh, 2017).

Algorithmic Design with Bias Mitigation

If algorithms are used (e.g., for content filtering, recommendation engines), design them to resist amplification of sensational content (countering availability bias) and the exposure of users to diverse viewpoints (countering confirmation bias), while also ensuring transparency where possible (Floridi, 2013).

Education & Literacy

Provide users with clear, accessible education on common cognitive biases and how they can affect blockchain interactions, empowering them to recognize their own biases and engage System 2 more effectively (Kahneman, 2011).

Blockchain's Potential to Influence Biases

While blockchain systems are susceptible to human biases, their unique characteristics also offer potential mechanisms to *counteract* certain biases, provided the systems are designed with these goals in mind and users engage deliberately (using System 2). It's crucial to recognize these are potentials, not guarantees, and require conscious effort in design and usage.

Transparency vs. Availability Bias

Blockchain's inherent transparency, by making historical transaction data immutable and publicly accessible, *can* mitigate availability bias *if* platforms provide tools for users to easily access and analyze this data (Xu & Livshits, 2018). Access to verifiable history allows for a more data-driven System 2 analysis, potentially countering the impact of sensationalized or easily available anecdotal evidence (Gorkhali & Chowdhury, 2022).

Immutability and Accountability

The immutability of records can foster accountability, which *could* counteract the tendency towards excessive coherence or loss aversion-driven cover-ups *if* platforms are designed to highlight and make these immutable records easily auditable and linkable to specific actions or decisions within a system (e.g., immutable DAO voting records) (Hsieh, 2017). This transparency can reinforce positive norms.

Decentralization and Groupthink

Decentralized governance, by distributing decision-making power, *has the potential* to reduce groupthink driven by centralized authority or excessive coherence. However, this requires conscious design of governance mechanisms to ensure diverse participation and information flow, rather than simply relying on token-weighted voting which can concentrate power and amplify the biases of a few large holders (Hsieh, 2017; Xu & Livshits, 2018).

The effectiveness of blockchain features in mitigating biases depends heavily on conscious design choices that facilitate System 2 engagement and provide access to relevant, understandable information, while also navigating the social dynamics that influence the adoption and interpretation of these features within the "imagined orders" of the blockchain community (Harari, 2014).

Culture, Motivations, and User Decision-Making

Understanding the interplay of culture, motivations, and individual decision-making patterns is crucial for designing user-centric blockchain solutions. While biases are universal cognitive tendencies, their expression and the most effective ways to mitigate them can vary based on cultural context, individual goals, and the specific user group (Hofstede, 2001).

Key factors influencing user engagement, viewed through a behavioral lens:

Perceived Value (Prospect Theory, Framing)

Users engage if they perceive a clear benefit relative to alternatives, framed in a way that resonates with their motivations (financial gain, control, community) (Kahneman, 2011; Kahneman & Tversky, 1979). Design must clearly articulate this value proposition, addressing potential loss aversion by highlighting potential gains and mitigating perceived risks through clear information.

Ease of Use (System 1/System 2, Cognitive Load)

Complex interfaces and technical requirements impose high cognitive load, discouraging users who default to System 1 (Kahneman, 2011). Prioritizing user-friendly UI/UX, guided onboarding, and minimizing jargon reduces this friction, making System 2 tasks (like setting up a wallet or understanding a transaction) less effortful. Research in human-computer interaction consistently shows the importance of usability for technology adoption (Neilsen, 1994).

Trust and Security (Availability Bias, Anchoring)

Trust in a decentralized system is built not just on cryptography but on user perception, which is heavily influenced by availability bias (news of hacks) and anchoring (initial impressions) (Gorkhali & Chowdhury, 2022; Kahneman, 2011; Kahneman & Tversky, 1974). Robust security design must be paired with transparent communication about risks and protective measures, helping users form a more accurate, less bias-driven assessment of trustworthiness. Harari's "imagined security" of the system relies on both technical reality and shared belief (Harari, 2014).

Ethics (Fairness, Trust, Social Norms)

Ethical considerations, informed by frameworks like the ACM Code of Ethics, must guide design (Floridi, 2013). This aligns with human needs for fairness and trust. Designing systems that prevent algorithmic bias, protect privacy, and ensure equitable access resonates with fundamental human values and builds a more sustainable, trusted "imagined order" (Harari, 2014). Community management must actively foster ethical norms and address behaviors driven by biases like excessive coherence or loss aversion (e.g., punishing scams fueled by availability bias).

Understanding the diverse motivations and challenges faced by different user types (Individuals, Businesses, Developers, Communities) – detailed in the previous section – allows for tailored behavioral interventions. For example, addressing the volatility challenge for individuals might involve framing strategies, while addressing the integration challenge for businesses might focus on reducing the perceived complexity and System 2 effort of adopting new technology.

Blockchain for Social Good: Leveraging Behavioral Insights

Behavioral science insights are not just for optimizing adoption or mitigating risks; they are essential for designing blockchain applications that actively drive positive social change, as explored with examples in Chapter 4.

Sustainable Practices (Nudging, Incentives)

Designing blockchain systems that track and incentivize environmentally friendly behavior (e.g., rewarding users for choosing renewable energy sources on a decentralized grid, verifying ethical sourcing in supply chains) directly applies behavioral economics principles like nudging and tokenized incentives to promote sustainable choices (Harari, 2014).

Financial Inclusion (Reducing Cognitive Barriers, Building Trust)

Applying behavioral insights to reduce the complexity and perceived risk of blockchain-based financial tools (microfinance, remittances) is crucial for reaching underserved populations. This involves simplifying interfaces, building trust through transparent processes, and framing benefits in culturally relevant ways, addressing the cognitive barriers and lack of trust often stemming from past experiences with traditional finance (Hofstede, 2001).

Community Empowerment (Behavioral Governance Design)

Using behavioral insights to design more inclusive, engaging, and equitable governance mechanisms in DAOs (as discussed earlier) can genuinely empower communities, ensuring participation is not hindered by cognitive load or social biases and that decisions reflect the diverse needs of members, building stronger, more resilient digital "imagined orders" (Harari, 2014; Hsieh, 2017).

The future of blockchain is in our hands. It requires collaboration between technologists, policymakers, behavioral scientists, social scientists, and the public. Integrating human-centered design and behavioral insights, informed by the macro perspective of shared beliefs (Harari) and the micro realities of human cognition (Kahneman), ensures blockchain technology serves humanity, creating a more equitable, sustainable, and informed digital future that aligns with our deepest needs and values.

The Indispensable Human Element

The indispensable human element of the theory of blockchain psychology lies in its fundamental recognition that blockchain technology, despite its mathematical and computational underpinnings, is ultimately a tool designed for and used by humans within social, cultural, and emotional contexts. It posits that the success, adoption, and ethical deployment of blockchain are not solely determined by its technical capabilities but are profoundly shaped by how humans perceive, interact with, and make decisions within blockchain-based systems.

Here is why the human element of blockchain technology is indispensable:

Adoption Drivers

Blockchain's potential can only be realized if individuals, businesses, and communities choose to adopt and utilize it. This adoption is driven by human motivations, perceived value, trust, and ease of use — all core areas of study within psychology and behavioral science. Without understanding what motivates people, why they trust (or distrust) new technologies, and what makes a system user-friendly, blockchain solutions risk remaining niche or failing to achieve their intended impact.

Navigating Biases

Human decision-making is riddled with cognitive biases. These biases significantly influence how individuals perceive and interact with blockchain, from investment decisions in cryptocurrencies (influenced by loss aversion or availability bias) to participation in decentralized governance (affected by confirmation bias or excessive coherence). Ignoring these inherent human tendencies in the design and implementation of blockchain systems can lead to unintended consequences, irrational behaviors, and ultimately hinder the technology's potential for positive change.

Designing User-Centric Solutions

For blockchain to be truly transformative, it must be designed with a deep understanding of human cognitive processes, emotional responses, social dynamics, and cultural contexts. User interfaces, communication strategies, and governance mechanisms that fail to align with these human factors will likely face resistance or be ineffective. The theory emphasizes "cognitive alignment" and "emotional resonance" as crucial design principles.

Building Trust and Security

Trust in decentralized systems is a complex psychological phenomenon. While blockchain's transparency and cryptography aim to build trust, user perception of security and the actual trustworthiness of the system are intertwined with emotional factors and cognitive biases. Behavioral insights can inform strategies to foster trust by addressing anxieties, simplifying complex security measures, and promoting transparent communication.

Shaping Ethical Outcomes

Blockchain technology is not inherently good or bad; its ethical implications are determined by how humans design and use it. The theory of blockchain psychology underscores the importance of considering human values, ethical frameworks, and potential societal impacts (like exacerbating inequalities or environmental concerns) during the development and deployment of blockchain solutions. Understanding human moral intuitions and social justice principles is crucial for guiding blockchain towards ethical applications.

Understanding Governance and Communities

Decentralized governance models (like DAOs) are fundamentally about human interaction and collective decision-making. Principles of social psychology, including group behavior, cooperation, conflict resolution, and the formation of social norms, are indispensable for designing effective and equitable governance mechanisms within these blockchain-based communities.

Cultural Context of Adoption

Blockchain adoption is not uniform across the globe. Cultural values, norms, and existing social structures significantly shape how different communities perceive and utilize the technology. Ignoring these cultural nuances can lead to the development of solutions that are irrelevant or even counterproductive in certain contexts.

In essence, the theory of blockchain psychology recognizes that blockchain is a socio-technical system. Its technical aspects are intertwined with the psychological and social realities of its users. By placing the human element at the forefront, the theory aims to move beyond a purely technical understanding of blockchain and provide a framework for creating blockchain solutions that are not only innovative and efficient but also user-friendly, trustworthy, ethical, and ultimately aligned with human needs and aspirations within diverse societal contexts. Without this indispensable human-centric perspective, blockchain risks failing to achieve its transformative potential and could even exacerbate existing societal problems.

Who Uses Blockchain?

There are several types of blockchain users, each with their own motivations, opportunities, and challenges:

Individuals

Motivations	Opportunities	Challenges
Financial Inclusion Access to financial services like microloans and remittances, especially in underserved communities.	**Financial empowerment** Access to new financial tools and opportunities.	**Navigating complexity** Understanding blockchain technology and managing security risks.
Data Ownership Control personal data and digital identity.	**Greater control over personal data** Enhanced privacy and security.	**Volatility of crypto markets** Potential for financial losses in investments.
Investment Participate in cryptocurrency and NFTs and other blockchain product markets.	**New forms of community and collaboration** Participation in DAOs and online communities.	**Scams and fraud** Identifying and avoiding malicious actors in the blockchain space.
Community Engagement Participate in decentralized governance like joining a DAO.		

Businesses

Motivations	Opportunities	Challenges
Efficiency and Transparency Streamline supply chains, track goods, and improve transparency and operational efficiency.	**Gain a competitive advantage** Improve efficiency, reduce costs, and enhance trust with customers.	**Integration with existing systems** Adapting blockchain technology to current infrastructure.
Reduce Costs Automate processes and eliminate intermediaries.	**New markets and revenue streams** Leverage blockchain for new business opportunities.	**Scalability and interoperability** Ensuring blockchain solutions can handle large-scale operations.
Innovation Explore new business models and create innovative products and services.		**Regulatory uncertainty** Navigating the evolving legal and regulatory landscape.
Improved Customer Experience Blockchain enables more quality time for businesses to improve their value proposition for their customers.		

Developers

Motivations	Opportunities	Challenges
Technological Innovation Build and improve blockchain platforms and applications.	**Shape the future of technology** Be at the forefront of blockchain innovation.	**Staying ahead of the curve** Keeping up with the rapid pace of blockchain development.
Solve Real-World Problems Create solutions that address behavioral, social and environmental challenges.	**Create impactful solutions** Develop applications that benefit society and the environment.	**Security and scalability** Building secure and scalable blockchain solutions.
Community Building Contribute to open-source projects and participate in blockchain communities.		**User Adoption** Designing user-friendly and accessible blockchain applications.

Communities

Motivations	Opportunities	Challenges
Decentralized Governance Create more democratic and transparent decision-making processes.	**Build stronger and more resilient communities** Foster trust and collaboration among members.	**Achieving consensus** Navigating diverse perspectives and making collective decisions.
Collective Ownership Manage shared resources and assets on the blockchain.	**Empower marginalized groups** Provide access to resources and opportunities in a more inclusive way.	**Security and privacy** Protecting community data and ensuring member privacy.
Social Impact Use blockchain to address social and environmental issues and causes.		**Scalability** Managing the growth and complexity of blockchain-based communities.

Understanding these diverse blockchain users and their motivations is crucial for developing effective solutions and promoting the responsible adoption of this transformative technology. Blockchain developers often focus on technical intricacies, but the human element is equally crucial. People are not purely rational; emotions, biases, and mental shortcuts shape our interactions with technology.

This chapter has delved into the indispensable human element at the heart of blockchain technology. We have moved beyond a purely technical understanding to explore how human cognition, biases, motivations, and social dynamics fundamentally shape the adoption, design, and impact of decentralized systems. Drawing on the foundational work of Daniel Kahneman and Amos Tversky, the historical perspective of Yuval Noah Harari, and contributions from broader behavioral and social science research, we have examined how predictable human behaviors manifest in the blockchain era – from loss aversion in volatile markets to confirmation bias in online communities.

We introduced the Theory of Blockchain Psychology as a framework for understanding this complex interplay, highlighting the critical importance of cognitive alignment, emotional resonance, social dynamics, cultural context, and ethical considerations in designing human-centered blockchain solutions. By recognizing that blockchain adoption is not purely rational, that human factors are predictable, and that technology and behavior are interdependent, this theory offers a lens for anticipating challenges and designing interventions.

Understanding the "human algorithm" is not just an academic exercise; it is essential for building a blockchain future that is truly beneficial for humanity. By acknowledging our biases, designing for our cognitive limitations, fostering positive social dynamics, and prioritizing ethical considerations, we can harness blockchain's power to create more intuitive, trustworthy, equitable, and sustainable systems. As we move forward, the insights from behavioral science will remain crucial tools for navigating the complexities of the decentralized web and ensuring that this transformative technology serves our collective well-being.

Chapter 4: Building a More Equitable & Sustainable World

In a world grappling with persistent inequalities and urgent environmental challenges, blockchain technology emerges as a potent, albeit complex, force for potential change. This chapter delves into how, in theory and with careful human-centered design, blockchain could contribute to dismantling existing power structures and fostering a more equitable and sustainable planet. We will explore its potential in areas like financial inclusion, land rights, transparent governance, and environmental stewardship. However, realizing this potential requires a critical understanding that technology is shaped by, and in turn shapes, human behavior and collective beliefs. We must approach blockchain's potential through the lenses of behavioral science and the dynamics of shared "imagined orders" (Harari, 2014), acknowledging blockchain's limitations, the complexities of achieving true social good, and the essential need for social, emotional, and cultural intelligence in its design and implementation (Di Ciccio, 2024).

A Human-Centered Approach to Equity: Navigating Biases and Building Trust

Addressing persistent global inequalities through technology is not solely a technical problem; it is fundamentally a human one, deeply influenced by psychology, social dynamics, and cultural contexts (Harari, 2014). While blockchain offers powerful tools to potentially level the playing field, its success hinges on designing and deploying solutions that account for how people actually behave, perceive risk, and form trust – insights provided by behavioral science (American Psychological Association, 2021; Iaccarino, 2023). Work on cognitive biases highlights the predictable "irrationalities" in human judgment (Ajzen, 1991), while the exploration of "imagined orders" reveals how shared beliefs structure our social realities and institutions, including those that perpetuate inequality (Harari, 2014).

Bridging the gap between the privileged and the marginalized with blockchain requires actively countering biases and building new, inclusive imagined orders that people can understand and trust. This necessitates integrating social, emotional, and cultural intelligence into blockchain design and implementation. These aren't just soft skills; they are practical applications of behavioral science insights in diverse human contexts.

Social Intelligence (Applying Behavioral Group Dynamics)

Understanding group behavior, power dynamics, communication styles, and resistance to change within different communities is vital (Bandura, 1977). Behavioral science teaches us about conformity, groupthink (amplified by confirmation bias), and how social norms influence technology adoption (Asch, 1956). Social intelligence applies these insights to design blockchain systems that foster inclusive participation, manage conflict constructively, and navigate existing social structures without causing unintended harm or alienation. It helps us understand, for instance, why a community might resist a transparent land registry system due to fear of disrupting existing, complex (but trusted within their "imagined order") informal systems (Thaler & Sunstein, 2008).

Emotional Intelligence (Applying Behavioral Principles of Trust and Motivation)

The capacity to recognize and manage emotions – fear of new technology, anxiety about financial risk (loss aversion), excitement (availability heuristic), resistance to change, trust formation – is crucial (Ajzen, 1991; Hofstede, 2001). Behavioral science provides insights into how emotions influence decision-making, how trust is built and broken, and how motivation is shaped by incentives and framing (Ajzen, 1991; Hofstede, 2001). Emotional intelligence uses this knowledge to design blockchain interfaces and communication strategies that build trust, address user anxieties directly, and frame the benefits in emotionally resonant ways, encouraging engagement despite psychological barriers.

Cultural Intelligence (Applying Behavioral Variation and Narrative Understanding)

Awareness and sensitivity to how cognitive biases, social norms, and communication styles manifest differently across cultures is essential (Rokeach, 1960). "Imagined orders" highlight how diverse cultural narratives shape different societies (Harari, 2014). Cultural intelligence applies behavioral insights within these diverse frameworks, ensuring blockchain solutions aren't based on Western-centric assumptions. It helps understand how cultural perceptions of privacy, ownership, risk, and community might influence the adoption of decentralized identity or governance, tailoring approaches to resonate with local values and existing narratives.

Blockchain Applications for Equity: Behavioral Hurdles and Human-Centered Design

Millions worldwide are excluded from traditional financial and governance systems. Blockchain offers potential solutions, as introduced in Chapter 1, but their success depends heavily on navigating behavioral hurdles and leveraging social, emotional, and cultural intelligence.

Financial Inclusion for the Underserved

Potential

As explored in Chapter 1, decentralized identity management, microfinance platforms, and remittance services built on blockchain can empower individuals lacking documentation or access to traditional banks (Hofstede, 1980). Research indicates that blockchain technology has the potential to improve financial inclusion by facilitating digital financial transactions, increasing financial savings, providing credit, and formalizing remittances (Hofstede, 1980; Iaccarino, 2023).

Behavioral Hurdles

Users face significant cognitive load in understanding novel blockchain concepts (wallets, keys, gas fees) (Ajzen, 1991; Nadler et al., 2024). Loss aversion makes potential users wary of putting limited resources into volatile cryptocurrencies or unfamiliar digital systems, especially if they have negative past experiences with informal finance (Ajzen, 1991). Availability bias can amplify fears due to widespread news of crypto scams (Ajzen, 1991). Overcoming resistance rooted in existing informal "imagined financial orders" requires building trust in the new digital one (Thaler & Sunstein, 2008).

Behavioral Science-Informed Design

Designing interfaces for cognitive ease with minimal jargon and intuitive steps is critical (Ajzen, 1991;

Nadler et al., 2024). Framing potential benefits (e.g., lower fees, faster transactions, secure savings) to counteract loss aversion is key (Ajzen, 1991). Implementing robust user support and education counters availability bias by providing reliable information (Ajzen, 1991). Cultural intelligence is needed to adapt communication and interfaces to local languages, digital literacy levels, and existing financial practices (Rokeach, 1960). Emotional intelligence helps build trust through transparent processes and responsive support, addressing user anxieties.

Land Rights and Property Management

Potential

Secure, transparent, immutable land registries, as introduced in Chapter 1, can protect individuals from land grabbing and empower owners, particularly in regions with weak governance (Mhlanga, 2022). Research on blockchain-based land registry projects in places like India suggests that digitizing land records with blockchain can amplify the benefits and avoid drawbacks of traditional systems (Mhlanga, 2022).

Behavioral Hurdles

Significant resistance to change rooted in deeply ingrained, traditional "imagined orders" of land ownership and power structures exists (Harari, 2014). Loss aversion is related to fearing the irreversible loss of ownership if the complex digital system is mishandled (Ajzen, 1991). Cognitive load is a factor in understanding digital registration processes (Ajzen,

1991). Building trust in a novel, abstract digital record requires overcoming availability bias related to past failures of centralized registries or digital systems (Ajzen, 1991).

Behavioral Science-Informed Design

Framing the digital registry benefits in terms of enhanced security and empowerment is crucial (Ajzen, 1991). Design must prioritize extreme user-friendliness and provide high-touch, culturally sensitive local support to reduce cognitive load (Ajzen, 1991; Nadler et al., 2024; Rokeach, 1960). Social intelligence is needed to engage with community leaders and address power dynamics resisting transparency (Bandura, 1977). Cultural intelligence is vital to respect existing land ownership narratives and integrate the system in a way that aligns with local customs and legal frameworks (Rokeach, 1960).

Supply Chain Transparency for Ethical Sourcing

Potential

As discussed in Chapter 1, tracking goods from origin to consumer increases transparency (Kshetri, 2021). For equity and sustainability, this means enabling ethical sourcing and combating issues like child labor or environmental damage in supply chains (Kshetri, 2021). Blockchain's ability to record every transaction allows for a transparent trail, ensuring authenticity and quality assurance (Kshetri, 2021).

Behavioral Hurdles

Lack of trust and willingness to share data between supply chain actors (influenced by past experiences and potential availability bias about misuse of data) is a challenge (Ajzen, 1991; Hofstede, 2001). Cognitive load is involved in integrating new tracking systems (Ajzen, 1991). Motivating participation from diverse, potentially less technologically sophisticated stakeholders is necessary. Resistance to transparency that might reveal unethical practices (loss aversion of profits) can occur (Ajzen, 1991).

Behavioral Science-Informed Design

Designing incentive structures (tokenized or otherwise) that leverage behavioral principles by framing participation as a gain (e.g., increased consumer trust, access to new markets) rather than just a cost or risk is important (Ajzen, 1991; Rokeach, 1960). Simplifying data input interfaces for cognitive ease is crucial (Ajzen, 1991; Nadler et al., 2024). Using social intelligence to build collaborative relationships and address power imbalances in the chain is necessary (Bandura, 1977). Cultural intelligence helps navigate different business norms and communication styles regarding data sharing and trust (Rokeach, 1960). The transparency itself helps build a shared, more trustworthy "imagined order" of product origin (Thaler & Sunstein, 2008).

Decentralized Governance (DAOs) for Community Empowerment

Potential

As introduced in Chapter 1 and discussed further in Chapter 3 regarding their structure, DAOs enable secure, transparent voting and participation, increasing accountability and community control over shared resources or decisions (Hofstede, 1980; Vendette & Thundiyil, 2023).

Behavioral Hurdles

Low participation due to cognitive load (understanding proposals) and decision fatigue is a challenge (Ajzen, 1991; Santana & Albareda, 2024). Power concentration through token-weighted voting can replicate existing hierarchies, leading to loss aversion regarding perceived influence for small holders (Ajzen, 1991; Nadler et al., 2024). Excessive coherence and confirmation bias can create echo chambers, undermining diverse input and leading to groupthink (Ajzen, 1991). Managing conflict in a decentralized, often pseudonymous, environment requires navigating complex social dynamics (Bandura, 1977). Research indicates that DAOs face significant challenges like declining participation, increasing centralization, and difficulties adapting (Schmitt, 2025).

Behavioral Science-Informed Design

Designing user interfaces that simplify proposals and voting processes for cognitive ease is important (Ajzen,

1991; Nadler et al., 2024). Implementing behavioral nudges or gamification can encourage participation. Exploring alternative governance models (e.g., quadratic voting) informed by social science can mitigate power concentration and encourage broader engagement (Bandura, 1977). Incorporating features that encourage exposure to diverse viewpoints and structured debate formats can counteract confirmation bias and excessive coherence (Ajzen, 1991). This requires high social and emotional intelligence to foster constructive communication and manage the inevitable conflicts that arise when challenging existing or forming new "imagined governance orders" (Hofstede, 2001; Thaler & Sunstein, 2008).

Real-World Examples: Behavioral Insights in Practice

Examining real-world implementations of blockchain technology through a behavioral lens reveals why some initiatives gain traction while others struggle, and showcases the diverse potential and inherent challenges in leveraging this technology for social good and sustainability. Moving beyond theoretical potential, these examples highlight blockchain's application across various sectors, demonstrating both successes and the complex human factors influencing adoption and impact, illustrating the behavioral hurdles and design considerations discussed above.

Kiva (Microfinance)

Kiva, a platform facilitating microloans, has explored blockchain for enhanced efficiency and transparency. Kiva's long-standing success, even pre-blockchain, relied heavily on emotional intelligence to build trust between lenders and borrowers through personal stories and transparent repayment tracking, countering the availability bias about the risks of lending to the poor (Ajzen, 1991; Hofstede, 2001). Blockchain integration, such as their pilot with Stellar for faster disbursements, enhances transparency (addressing availability bias) and potentially reduces cognitive load by streamlining processes (Ajzen, 1991), but still requires cultural intelligence to adapt financial concepts and communication to diverse local contexts and social intelligence to facilitate the peer-to-peer connection that fosters trust in a digital environment (Bandura, 1977; Mhlanga, 2022; Thaler & Sunstein, 2008), illustrating the behavioral considerations for financial inclusion (Hofstede, 1980).

UN World Food Programme (WFP) Building Blocks

The WFP utilizes a private blockchain, Building Blocks, for humanitarian aid delivery, specifically cash transfers to refugees in Jordan and Bangladesh. This system records and verifies transactions, allowing aid recipients to receive and spend assistance securely in local markets. Success depends on reducing the cognitive load for recipients interacting with new technology through user-friendly interfaces (Ajzen,

1991), building trust in the digital system (countering availability bias related to past aid mismanagement or digital scams) (Ajzen, 1991; Hofstede, 2001) and navigating complex local social structures and power dynamics (social intelligence) (Bandura, 1977; Rokeach, 1960). The system aims to create a transparent, trusted "imagined order" for aid distribution (Thaler & Sunstein, 2008), with reported benefits including reduced costs, increased security, and greater financial inclusion for recipients, demonstrating how behavioral insights are crucial for effective aid delivery via blockchain (Hofstede, 1980).

Provenance (Supply Chain Transparency for Ethics)

Provenance leverages blockchain to create transparent and traceable supply chains, allowing consumers to track the origin and journey of products, such as fish or clothing. This aims to build consumer trust in ethical sourcing and sustainability claims. Research explores the role of blockchain in enabling end-to-end traceability and data integrity in supply chains for sustainable sourcing (Kshetri, 2021). Success hinges on communicating this transparency effectively (framing effect, social intelligence in presenting complex data) (Ajzen, 1991; Bandura, 1977) and presenting complex supply chain information in a way that is easily digestible (cognitive ease) (Ajzen, 1991), countering consumer skepticism potentially fueled by availability bias about past greenwashing or unethical sourcing narratives (Ajzen,

1991). Challenges include motivating all actors in a complex supply chain to participate and input accurate data, requiring behavioral strategies to build trust and incentivize participation (Hofstede, 2001; Rokeach, 1960), highlighting the behavioral challenges in achieving supply chain transparency for ethical outcomes.

Follow My Vote (Online Voting Systems & Governance)

Follow My Vote is an open-source project exploring blockchain for secure and transparent online voting. The goal is to increase voter turnout and trust in election outcomes by providing a publicly auditable ledger of votes. Adoption faces significant behavioral hurdles related to trust in the technology itself (countering availability bias from news of hacks or system failures) (Ajzen, 1991; Hofstede, 2001), overcoming voter inertia (cognitive load of learning a new system) (Ajzen, 1991), and navigating political resistance rooted in existing "imagined orders" of electoral processes (Harari, 2014). While the technology offers potential for transparency, success depends on building widespread public confidence and designing interfaces for maximum cognitive ease and accessibility (Ajzen, 1991; Nadler et al., 2024), illustrating the behavioral challenges in implementing blockchain for democratic processes (Hofstede, 2001; Nadler et al., 2024).

Medicalchain (Healthcare Data Management & Equity)

Medicalchain uses blockchain to allow patients to control access to their electronic health records, enabling secure sharing with healthcare providers. This aims to improve interoperability and patient agency over personal health information, contributing to more equitable access to and control over health data. Research highlights the potential of blockchain to enhance privacy, security, and user control in digital identity management, shifting control from centralized authorities to individuals (Iaccarino, 2023). Implementation involves addressing patient anxieties about data security and privacy (emotional intelligence) (Ajzen, 1991; Hofstede, 2001), building trust among various stakeholders (patients, doctors, hospitals, insurers) who have different levels of digital literacy and trust in new systems (social intelligence) (Bandura, 1977; Hofstede, 2001), and navigating complex regulatory environments around health data (cultural intelligence regarding privacy norms) (Nelson & Charlotte, 2025; Rokeach, 1960). The system challenges existing "imagined orders" of data ownership and access in healthcare (Harari, 2014), highlighting the behavioral and social complexities of empowering individuals with control over sensitive data.

BitPesa (Cross-Border Payments & Financial Inclusion)

Now part of AZA Finance, BitPesa utilized blockchain to facilitate faster and cheaper cross-border payments, particularly in Africa. This addresses a significant barrier to financial inclusion and economic development (Hofstede, 1980; Iaccarino, 2023). Success required overcoming user hesitancy towards cryptocurrency volatility (loss aversion, emotional intelligence) (Ajzen, 1991; Hofstede, 2001), building trust in a new financial service (availability bias from past financial scams) (Ajzen, 1991; Hofstede, 2001), and adapting the service to diverse mobile money ecosystems and regulatory environments across different countries (cultural intelligence) (Nelson & Charlotte, 2025; Rokeach, 1960). The platform aimed to build a more efficient "imagined financial order" for remittances and business payments (Thaler & Sunstein, 2008), demonstrating the behavioral considerations for financial inclusion solutions (Hofstede, 1980).

Decentralized Energy Grids (e.g., Power Ledger Pilots & Sustainability)

Projects like Power Ledger have piloted blockchain-based platforms enabling peer-to-peer trading of renewable energy within local microgrids. This incentivizes renewable energy generation and more efficient energy use, contributing to sustainability (Sun et al., 2025). Adoption depends on motivating prosumers (producers/consumers) to participate (incentives based on behavioral principles) (Ajzen, 1991; Sun et al., 2025), simplifying the complex process of energy trading (cognitive ease) (Ajzen, 1991), building trust in the automated trading system (availability bias about energy grid reliability) (Ajzen, 1991; Hofstede, 2001), and navigating existing energy market regulations and consumer behaviors (social and cultural intelligence) (Bandura, 1977; Nelson & Charlotte, 2025; Rokeach, 1960). The initiative seeks to build a new "imagined order" for energy markets (Harari, 2014), empowering individuals and illustrating behavioral strategies for promoting sustainable actions (Sun et al., 2025). Research indicates that blockchain technology positively moderates the relationship between critical success factors and project success criteria of renewable energy[1] projects (Kowalski & Esposito, 2023).

Civic (Decentralized Identity Verification & Equity)

Civic provides a decentralized identity verification service using blockchain to allow users to control and reuse their verified personal information. This aims to reduce the need for repetitive KYC processes and enhance privacy, contributing to more equitable and privacy-preserving digital interactions. Adoption requires users to trust a decentralized method with sensitive data (countering availability bias related to identity theft) (Ajzen, 1991; Hofstede, 2001), understand and manage their digital keys (cognitive load) (Ajzen, 1991), and for businesses to accept this new form of identity verification (challenging existing verification "imagined orders") (Harari, 2014). Designing for cognitive ease and clearly communicating the security and privacy benefits (framing effect) are crucial (Ajzen, 1991), highlighting the behavioral challenges in shifting identity paradigms. Research explores how blockchain's features like encryption, decentralization, and cryptographic proofs enhance privacy and user control in digital identity management (Iaccarino, 2023).

These examples demonstrate that while blockchain technology provides powerful capabilities for social good and sustainability, its real-world impact is inextricably linked to human behavior. Successful implementations require not only technical robustness but also a deep understanding of psychological biases, social dynamics, and cultural contexts to build trust, encourage adoption, and navigate the complex process of challenging and building new "imagined orders" that can lead to a more equitable and sustainable future (American Psychological Association, 2021; Bandura, 1977; Harari, 2014; Rokeach, 1960).

Blockchain & Sustainability: Behavioral Drivers for a Greener Future

Addressing climate change and promoting sustainability requires large-scale behavioral shifts and collective action. Blockchain can be a powerful tool for this, but only if designed with an understanding of human motivation, biases, and the dynamics of collective action (Ajzen, 1991; Bandura, 1977; Thaler & Sunstein, 2008). Building a sustainable future via blockchain requires navigating existing "imagined orders" related to resource consumption and environmental responsibility and building new ones that incentivize green behavior (Harari, 2014; Sun et al., 2025). However, a balanced perspective necessitates confronting blockchain's own environmental footprint, particularly the energy consumption of certain protocols.

Blockchain's Environmental Impact: The Energy Debate

While blockchain offers promising applications for sustainability, it's crucial to address its own environmental cost. The primary concern revolves around the energy consumption of certain consensus mechanisms, most notably Proof-of-Work (PoW) (Conway, 2022). As discussed in Chapter 1, PoW requires miners to expend significant computational power to solve complex cryptographic puzzles. This competition consumes vast amounts of electricity in the real world. The security of a PoW network is directly tied to this energy expenditure – the more energy consumed, the more expensive a 51% attack becomes (OSL, 2025b). Large PoW networks, like Bitcoin, consume energy comparable to that of small to medium-sized countries. This high energy demand, particularly if powered by fossil fuels, contributes to carbon emissions and environmental concerns. The trade-off in PoW is clear: maximum decentralization and robust security (proven over time) are achieved at the cost of a significant environmental footprint (Conway, 2022).

Energy-Efficient Alternatives (PoS and Beyond)

The blockchain ecosystem is actively developing and transitioning towards more sustainable alternatives. PoS mechanisms replace computational work with economic stake. Validators are chosen based on the amount of cryptocurrency they hold and are rewarded for honest participation while risking their stake if they act maliciously. This process consumes dramatically less energy than PoW – typically orders of magnitude lower, comparable to running a standard computer server (Conway, 2022).

Other consensus mechanisms like Delegated Proof-of-Stake (DPoS), Proof-of-Authority (PoA), and various Byzantine Fault Tolerance (BFT) variants are also significantly more energy-efficient than PoW, relying on different models of trust and validation that do not require massive computational races (GeeksforGeeks, 2025; Roy, 2023). While vastly more energy-efficient, transitioning to or choosing these mechanisms involves different trade-offs, as discussed previously, such as potential shifts in centralization risk, different security assumptions, and relative novelty compared to PoW's long track record. However, from a purely environmental perspective, their advantage over PoW is substantial.

The Human Factor in the Energy Transition

The shift towards greener blockchain protocols is not purely a technical upgrade; it's a behavioral and collective action challenge. Adopting more sustainable mechanisms requires:

- Overcoming loss aversion from entities heavily invested in existing PoW mining hardware and infrastructure (Ajzen, 1991).

- Navigating the cognitive load for developers, validators, and users in understanding and trusting new, more complex consensus models (Ajzen, 1991).

- Addressing confirmation bias and resistance to change within communities deeply entrenched in the narratives and practices of PoW (Ajzen, 1991).

- Building collective trust (a new "imagined order") in the security and reliability of PoS or other efficient systems (Hofstede, 2001; Thaler & Sunstein, 2008).

Behavioral nudges, clear communication framing the long-term environmental and economic benefits, and designing energy-efficient options for cognitive ease can help drive this human-driven transition towards a more sustainable blockchain infrastructure (Ajzen, 1991; Hofstede, 2001).

Revolutionizing Resource Management & Traceability (Leveraging Blockchain for Environmental Good)

Beyond its own footprint, blockchain's characteristics can be leveraged to promote environmental sustainability. This could include transparent supply chains, ethical sourcing, circular economy tracking (Kshetri, 2021). Blockchain can support supply chain governance for social, environmental, and economic sustainability goals (Kshetri, 2021).

Motivating businesses to adopt transparency when it might reveal costly practices (loss aversion) is a challenge (Ajzen, 1991). Ensuring accurate data input despite the cognitive load of new systems and potential incentives for dishonesty is necessary (Ajzen, 1991). Building consumer trust in transparent labels despite availability bias from past greenwashing scandals is crucial (Ajzen, 1991). This entails designing incentives informed by behavioral principles, simplifying data input (cognitive ease) (Ajzen, 1991), and communicating transparency benefits effectively (framing effect) (Ajzen, 1991). Social intelligence is needed to foster collaboration across complex value chains (Bandura, 1977); cultural intelligence to adapt approaches to local resource use norms and business practices (Rokeach, 1960). This effort builds a shared, verifiable "imagined order" of sustainable resource flow (Thaler & Sunstein, 2008).

Tracking and Verifying Sustainability Initiatives (Building Trust and Credibility)

Blockchain's ability to securely and transparently record data makes it a potential tool for tracking and verifying sustainability efforts. This can include carbon footprint tracking, renewable energy trading, and sustainability certifications (Kowalski & Esposito, 2023; Kshetri, 2021; Sun et al., 2025. A blockchain-based solution for tracking mission-critical data in a value chain can provide a full provenance of a product or service's total emissions in near real-time (Kshetri, 2021).

Motivating individuals/businesses to accurately report data (incentives, trust) is necessary (Hofstede, 2001; Sun et al., 2025). Building trust in digital certificates or carbon credits despite availability bias from past fraudulent schemes is crucial (Ajzen, 1991; Hofstede, 2001). Encouraging participation in new markets (renewable energy trading) despite loss aversion or perceived complexity is a challenge (Ajzen, 1991). This requires designing robust, auditable systems that build trust by countering availability bias (Ajzen, 1991; Hofstede, 2001). Using framing and incentives based on behavioral principles can encourage data sharing and participation (Ajzen, 1991; Sun et al., 2025). Social intelligence is needed to build communities around initiatives (Bandura, 1977); cultural intelligence to adapt communication on environmental responsibility (Rokeach, 1960).

Incentivizing Green Behavior (Designing Effective Systems)

Blockchain can potentially create powerful incentive mechanisms to encourage sustainable practices with carbon credits, reward systems for eco-friendly actions (Sun et al., 2025). Blockchain technology could provide new ways to incentivize behavior of resource users (Sun et al., 2025).

Designing incentives that truly motivate, beyond short-term gain, is a challenge. Behavioral science principles are vital, framing rewards as gains and penalties as losses, and understanding how the magnitude and probability of outcomes influence behavior (Ajzen, 1991). Avoiding unintended consequences where incentives lead to gaming the system is necessary. This effort requires deep understanding of human motivation, social intelligence to gauge community response (Bandura, 1977), and cultural intelligence to adapt incentives to local economic and social norms (Rokeach, 1960). Designing these systems correctly helps build a new "effective choice architecture" where sustainable behavior is valued and rewarded (Thaler & Sunstein, 2008).

Navigating The Path to A Greener Blockchain: A Holistic Approach

Achieving a sustainable future requires a multifaceted approach that addresses both the environmental impact of blockchain technology itself and leverages its potential for environmental good. The urgency of addressing the energy consumption of protocols like PoW is paramount, and the ongoing human-driven transition to more efficient mechanisms like PoS is a critical step (Hofstede, 1980).

Blockchain's potential for social good and environmental stewardship is immense, but it is not automatic. It is contingent upon a conscious, human-centered design approach that leverages the insights of behavioral science to understand and positively influence user behavior (American Psychological Association, 2021; Ajzen, 1991), builds trust and facilitates collective action within new "imagined orders" (Bandura, 1977; Harari, 2014; Hofstede, 2001), and integrates social, emotional, and cultural intelligence at every step (Bandura, 1977; Hofstede, 1980; Rokeach, 1960). By focusing on how humans interact with and perceive these systems, and by actively addressing the technology's own environmental footprint, we can build blockchain solutions that are not only technically sound but also genuinely equitable, inclusive, and sustainable, forging a future where technology serves humanity's greatest needs.

Chapter 5: The Convergence of Blockchain and AI

Blockchain and Artificial Intelligence (AI), two of the most transformative technologies of our era, are increasingly converging, creating a powerful synergy with the potential to revolutionize diverse industries and reshape aspects of our lives. This chapter explores the dynamic relationship between blockchain and AI, examining how they mutually enhance each other's capabilities and the opportunities and challenges presented by their integration.

Enhancing AI Capabilities with Blockchain

Blockchain's inherent characteristics can significantly address some of the limitations and challenges associated with Artificial Intelligence, particularly in areas concerning data and trust.

One critical aspect of effective AI development is access to secure, reliable, and unbiased training data. Blockchain's decentralized and immutable ledger provides a robust platform for storing and managing AI training data, helping to ensure data integrity, provenance, and auditability (Zafar, 2025). This mitigates the risk of data manipulation or bias being introduced into AI algorithms, a significant concern in developing fair and trustworthy AI systems (Akter et al., 2025; Barik et al., 2023). By recording AI model training processes and updates on the blockchain, developers can create an immutable audit trail, increasing trust and accountability in AI systems (Samson & Williams, 2025). This transparency is particularly important in sensitive domains such as healthcare and finance, where the explainability and reliability of AI decisions are crucial (Kan, 2024). Empirical studies have demonstrated that integrating blockchain with AI in healthcare data security systems can lead to significant improvements in data protection, including a substantial reduction in unauthorized access attempts (Akter et al., 2025; Samson & Williams, 2025; Zafar, 2025).

Managing and sharing large datasets for AI applications, especially across multiple parties, presents considerable challenges related to privacy, security, and interoperability. Blockchain, with its distributed ledger technology, can facilitate secure and transparent data sharing and collaboration without relying on a central authority (Kan, 2024). This enables the development of more robust and comprehensive AI models that can be trained on diverse datasets while maintaining data privacy and control (Barik et al., 2023; Kan, 2024; Lee et al., 2013). This capability is particularly valuable in fields like healthcare, where stringent data privacy and security regulations are paramount, allowing researchers to access and analyze large, distributed datasets securely (Akter et al., 2025; Barik et al., 2023; Di Ciccio, 2024; Dixit & Jangid, 2025).

A notable challenge in AI is the "black box" problem, where the decision-making process of complex algorithms can be opaque and difficult to understand (Akter et al., 2025). Blockchain can contribute to addressing this by providing a transparent and verifiable record of key aspects of the AI lifecycle, including data sources used for training, model versions, and the logic applied in specific decisions (Akter et al., 2025; Barik et al., 2023). This increased transparency, stored immutably on the blockchain, can foster greater trust in AI systems and enable better understanding and interpretation of AI-driven outcomes by users and stakeholders (Akter et al., 2025). A framework leveraging blockchain for AI transparency has been proposed to improve decision traceability, data provenance, and model accountability (Akter et al., 2025).

Enhancing Blockchain Capabilities with AI

Conversely, Artificial Intelligence can bring significant advancements to blockchain technology, improving its efficiency, scalability, and security.

Integrating AI into smart contracts can create more sophisticated, adaptable, and autonomous agreements (Dixit & Jangid, 2025; Patel, 2024; Samson & Williams, 2025; Trigsted, 2025). AI algorithms can analyze real-world data streams and trigger actions based on complex, predefined conditions embedded within the smart contract code, automating intricate processes and potentially reducing the need for human intervention (Patel, 2024; Samson & Williams, 2025). This convergence can lead to more efficient and dynamic smart contracts applicable in various sectors, including supply chain management, insurance, and decentralized finance (DeFi) (Samson & Williams, 2025). Research explores how combining blockchain, smart contracts, and machine learning can address challenges in fintech related to security, transparency, and operational performance (Dixit & Jangid, 2025; Roberts et al., 2021; Patel, 2024).

Scalability remains a significant challenge for many blockchain networks, limiting their transaction throughput and speed (Dixit & Jangid, 2025). AI techniques can offer innovative solutions to enhance blockchain scalability by optimizing various operational aspects (Dixit & Jangid, 2025; Patel, 2024). This can include using AI for predictive analytics to optimize transaction routing and manage network traffic, identifying and mitigating bottlenecks, and potentially enhancing consensus mechanisms to be more efficient (Dixit & Jangid, 2025; Samson & Williams, 2025; Finance Technology Insights, 2024; Roberts et al., 2021). Empirical evaluations are exploring how AI techniques can improve measurements related to efficiency and performance in blockchain systems (Finance Technology Insights, 2024; Roberts et al., 2021).

AI can significantly bolster the security of blockchain networks. By analyzing the vast datasets recorded on the blockchain, AI algorithms can identify suspicious patterns, anomalies, and potentially fraudulent activities in real-time with greater accuracy than traditional methods (Barik et al., 2023; Patel, 2024). AI-powered fraud detection systems can reduce false positives and improve the identification of illicit transactions (Barik et al., 2023). The convergence of AI with blockchain's immutable record-keeping further strengthens fraud prevention and can enable smart contract-based compliance enforcement (Barik et al., 2023; Kan, 2024).

Blockchain networks accumulate large amounts of data over time, presenting opportunities for advanced data analytics. AI can be used to analyze these extensive datasets, providing valuable insights into network patterns, usage trends, and potential anomalies (Barik et al., 2023; Paul & Ogburie, 2025; Roberts et al., 2021; Samson & Williams, 2025). This AI-driven analysis can help businesses and organizations make more informed decisions, optimize their operations when using blockchain solutions, and gain deeper insights from the transparent ledger (Paul & Ogburie, 2025). Combining AI analytics with blockchain technology provides solutions for enhanced business intelligence, addressing limitations in traditional systems related to security and data quality (Samson & Williams, 2025; Paul & Ogburie, 2025).

Protecting Digital Assets: Likeness and Intellectual Property

The convergence of blockchain and AI also presents crucial considerations and potential solutions for protecting individual likeness and intellectual property in the digital age, especially with the rise of AI-generated content and synthetic media.

With AI capable of generating realistic digital likenesses, managing consent and usage rights is paramount. Blockchain can empower individuals with greater control over their digital likenesses through decentralized identity and consent management systems (Flatworld Solutions, 2025; Lee et al., 2013; Nelson & Charlotte, 2025; Trigsted, 2025; Zafar, 2025). By creating a secure and decentralized identity linked to blockchain, individuals can manage and explicitly grant permission for the use of their likeness in AI applications, potentially ensuring fair compensation and the right to refuse usage they do not agree with (Flatworld Solutions, 2025; Lee et al., 2013; Trigsted, 2025). Blockchain can create an immutable audit trail of every transaction and interaction involving an individual's likeness, which can be used to track usage and ensure it aligns with consent (Flatworld Solutions, 2025; Nelson & Charlotte, 2025; Trigsted, 2025). Smart contracts can further automate the enforcement of likeness rights agreements, programming automatic payments when a likeness is used, thereby ensuring fair compensation and reducing disputes (Flatworld Solutions, 2025; Trigsted, 2025).

AI models are often trained on vast amounts of data, including creative works and intellectual property. Blockchain can be used to establish clear and tamper-proof proof of ownership and provenance for intellectual property used in AI development or generated by AI (Garcia et al., 2025). This can help prevent copyright infringement and ensure creators are fairly compensated (Garcia et al., 2025). Blockchain can also facilitate secure and transparent data sharing and collaboration among AI developers and researchers while protecting underlying intellectual property rights through granular permissions and immutable records (Garcia et al., 2025; Kan, 2024).

Use Cases

AI-Generated Art and Music

Blockchain can track the ownership and provenance of AI-generated art and music, protecting the rights of AI developers and individuals whose data or likenesses were used (Garcia et al., 2025; Nelson & Charlotte, 2025; Paul & Ogburie, 2025; Trigsted, 2025).

Synthetic Media and Deepfakes

Blockchain can help mitigate risks associated with deepfakes and synthetic media by providing a transparent record of the origin and manipulation of digital content, helping verify authenticity and prevent the spread of misinformation (Garcia et al., 2025; Paul & Ogburie, 2025).

AI-Powered Content Creation

Blockchain can protect the intellectual property rights of content creators using AI tools by recording AI model and data usage on the blockchain, establishing proof of ownership and preventing unauthorized use of their work and source material (Garcia et al., 2025; Paul & Ogburie, 2025). Empirical studies are reviewing how blockchain technologies provide opportunities for media copyright management, identifying areas like Digital Rights Management and Intellectual Property Rights (Garcia et al., 2025; Paul & Ogburie, 2025).

Challenges and Considerations

Despite the immense potential, the convergence of blockchain and AI is not without its challenges.

Scalability and Computational Overhead

Integrating blockchain and AI can present scalability challenges, as both technologies can be computationally intensive. AI applications requiring real-time data processing may be hindered by blockchain's transaction throughput limitations (Dixit & Jangid, 2025; Lee et al., 2013). The computational overhead of combining these technologies needs to be addressed for widespread adoption (Dixit & Jangid, 2025; Lee et al., 2013; Trigsted, 2025). Research highlights scalability as a significant challenge in the widespread adoption of blockchain technology, with traditional blockchains struggling to scale with increasing transaction demands (Samson & Williams, 2025; Trigsted, 2025; Dixit & Jangid, 2025).

Interoperability

Ensuring seamless interoperability between different blockchain platforms and between blockchain and AI systems can be complex (Nelson & Charlotte, 2025). The absence of standardized protocols and the diversity of existing systems pose technical hurdles (Nelson & Charlotte, 2025).

Ethical Considerations

The combination of blockchain and AI raises significant ethical considerations. These include the potential for biased algorithms trained on prejudiced data stored on blockchain, data privacy concerns when combining transparent ledgers with powerful analytical AI, and the impact on human employment as automation increases (Barik et al., 2023). Leveraging blockchain for ethical AI development by emphasizing trust, transparency, and fairness is an opportunity (Barik et al., 2023).

Regulatory and Governance Challenges

The decentralized nature of blockchain combined with the complexity of AI decision-making presents challenges for existing regulatory frameworks (Zafar, 2025). Establishing clear guidelines for accountability, liability, and compliance in integrated blockchain-AI systems is an ongoing process (Kan, 2024). Research explores the complex interplay between blockchain technology and data protection regulations, highlighting challenges and potential solutions for compliance in blockchain implementations (Kan, 2024).

Towards a Human-Centered AI and Blockchain Future

The convergence of blockchain, AI, and behavioral science presents a unique opportunity to create technological solutions that are not only powerful and efficient but also human-centered and ethical.

Behavioral Insights for Design

Behavioral science can inform the design and implementation of AI systems integrated with blockchain to align with human needs and values (Borghoff, 2025). Understanding human behavior, motivations, and biases can guide the creation of AI and blockchain solutions that are more user-friendly, trustworthy, and ethical (Borghoff, 2025; Dixit & Jangid, 2025; Roberts et al., 2021). Research explores the psychological factors that influence technology adoption decisions, which are relevant to user acceptance of integrated systems (Lee et al., 2013).

Explainable and Transparent Systems

Behavioral science can help address the "black box" problem by promoting transparency and explainability in AI-blockchain systems (Akter et al., 2025). Understanding how humans perceive and interact with these systems can guide the development of AI that is more understandable and trustworthy, fostering user confidence (Akter et al., 2025; Barik et al., 2023; Paul & Ogburie, 2025).

AI for Social Good

The combination of AI and blockchain can be leveraged to address social and environmental challenges (Nelson & Charlotte, 2025). By understanding human behavior and leveraging blockchain's secure and transparent infrastructure, AI can be used to promote sustainable practices, foster financial inclusion, and empower communities

(Borghoff, 2025; Nelson & Charlotte, 2025; Omar et al., 2025).

Integrating Human Intelligence

Designing AI systems to better understand and respond to social and emotional cues can lead to more natural and effective human-computer interactions within blockchain-based applications. Developing AI with cultural sensitivity ensures inclusivity and respect for diverse norms.

The integration of blockchain and AI within the context of Web 3.0 can create a more decentralized, user-centric, and intelligent internet (Paul & Ogburie, 2025). This can empower individuals with greater control over their data and online identities, foster more democratic and transparent online communities, and promote a more equitable and accessible digital landscape (Lee et al., 2013; Paul & Ogburie, 2025).

Specific use cases leveraging this convergence include enhancing Decentralized Autonomous Organizations (DAOs) with AI for more efficient decision-making, creating personalized education and healthcare solutions, and optimizing supply chain management with real-time data analysis and automation (Patel, 2024; Samson & Williams, 2025; Trigsted, 2025).

The convergence of blockchain, AI, and behavioral science offers a unique opportunity to create technological solutions that are not only powerful and efficient but also human-centered and ethical. By understanding human behavior and leveraging the strengths of both blockchain and AI, we can unlock the full potential of these technologies to address real-world challenges and build a more equitable and sustainable global digital community.

Chapter 6: The Future of Blockchain: Scenarios, Challenges, and Responsible Development

Shaping Blockchain's Future

Blockchain technology, having emerged from its historical roots and demonstrated diverse applications, stands at a pivotal juncture. Its future trajectory is not predetermined; rather, it is actively shaped by human choices, evolving societal values, and our collective adherence to shared beliefs – what Yuval Noah Harari terms "imagined orders" in *Sapiens* (Harari, 2014). This chapter delves into potential future scenarios where blockchain, informed by insights from behavioral science and other disciplines, could contribute to fostering a more equitable and sustainable world. However, realizing this transformative potential necessitates confronting significant technical, social, and ethical challenges (Santana & Albareda, 2024; Vendette & Thundiyil, 2023). We must critically examine blockchain's inherent limitations alongside its widely discussed benefits, understanding that technology develops within, and subsequently influences, our deeply ingrained social structures and "imagined orders" (Harari, 2014). This chapter aims to provide a balanced perspective on blockchain's promise and the critical prudence required to navigate its development responsibly, emphasizing the human element at the heart of this technological evolution.

Scenario 1: Decentralized Trust and Individual Empowerment

One compelling future scenario envisions blockchain fostering unprecedented levels of individual control and enabling more trustworthy interactions across digital domains. However, achieving this vision is contingent upon overcoming significant technical and behavioral hurdles.

Decentralized Identity: Promise and Friction

Decentralized identity management, often referred to as Self-Sovereign Identity (SSI), represents a paradigm shift with the potential to empower individuals with granular control over their digital selves, allowing secure and selective sharing of personal information without reliance on centralized authorities (Nelson & Charlotte, 2025). This approach aims to foster greater trust in online interactions by giving individuals agency over their data. Despite this promise, achieving user-friendly and widely adopted decentralized identity solutions remains a significant challenge (Chan et al., 2025). Hurdles include the technical complexity for end-users in managing cryptographic keys and understanding nuanced consent mechanisms. Ensuring seamless interoperability between diverse SSI systems and mitigating risks like irreversible identity loss in case of mishandling are also critical concerns.

The cognitive effort ("System 2" thinking) required to understand and interact with novel decentralized identity concepts often clashes with the ease and familiarity of traditional centralized login systems, which rely on intuitive "System 1" processing (Kahneman, 2011). This cognitive friction creates a significant barrier to widespread adoption. Furthermore, managing one's digital identity involves complex privacy and security risks that users might underestimate due to cognitive biases like the availability heuristic, where individuals may downplay risks they haven't personally experienced or seen widely reported (Tversky & Kahneman, 1973). Ultimately, the widespread adoption and trust in decentralized identity rely on a shared belief in the system's reliability and security – an "imagined order" of digital trust – a belief susceptible to erosion by both technical failures and these inherent cognitive biases (Harari, 2014).

Behavioral Nudges: Guidance vs. Manipulation

Blockchain platforms and decentralized applications (dApps) could potentially incorporate insights from behavioral science to design incentive structures and "nudges" that encourage desirable actions, such as promoting energy conservation on a decentralized energy grid or fostering participation in community governance (Thaler & Sunstein, 2008). These nudges aim to gently steer user behavior towards beneficial outcomes without restricting choice.

However, designing ethical nudges within blockchain systems requires careful consideration of user autonomy and the potential for manipulation (Nadler et al., 2024). Transparency regarding how nudges are implemented and ensuring explicit user consent are paramount, especially given the risk that opaque or poorly designed mechanisms could exploit inherent cognitive biases, leading users to make decisions that are not in their best interest (Kahneman, 2011). Nudges must be perceived by users as helpful guidance rather than coercive tactics, aligning with the understanding that powerful narratives and perceived intentions significantly shape behavior and trust within a shared "imagined order" (Harari, 2014).

Algorithmic Transparency: Ideal vs. Reality

Blockchain's inherent transparency, particularly in recording transactions and potentially the logic of smart contracts, could theoretically extend to making the algorithms that power integrated AI or govern decentralized systems more open to scrutiny. This transparency could enable collaboration between behavioral scientists, ethicists, and technical experts to identify and mitigate biases, promoting fairer decision-making processes within blockchain applications.

Despite the ideal of complete algorithmic transparency, ensuring genuine fairness and preventing subtle biases requires ongoing interdisciplinary collaboration and robust auditing mechanisms. The increasing complexity of AI models integrated with blockchain can complicate the detection and understanding of how decisions are made and where bias might be embedded. Furthermore, users often rely on intuitive "System 1" thinking to interact with technology, implicitly trusting algorithms without engaging in critical evaluation ("System 2") (Kahneman, 2011). This can potentially lead to the embedding of new "imagined hierarchies" within digital systems, perpetuating or even amplifying existing societal discrimination if algorithmic biases are not actively identified and addressed through continuous auditing and diverse input (Nelson & Charlotte, 2025).

Scenario 2: Community Empowerment and Economic Inclusion

Another significant future scenario revolves around blockchain's potential to empower communities and foster greater economic inclusion, particularly for underserved populations. However, realizing this requires navigating complex governance structures and addressing accessibility challenges.

Decentralized Governance: Participation and Pitfalls

Blockchain technology can enable secure and transparent voting systems, which could potentially increase participation and accountability in community decision-making processes, such as those within Decentralized Autonomous Organizations (DAOs) (Santana & Albareda, 2024). This offers a vision of more democratic and inclusive governance structures. Yet, effective decentralized governance faces substantial hurdles in practice (Zafar, 2025). Challenges include consistently low voter turnout in many DAO initiatives, the potential for disproportionate influence by large token holders or 'whales,' difficulties in resolving disputes within decentralized frameworks, and inherent security vulnerabilities within smart contracts and governance mechanisms (Zafar, 2025).

The cognitive load associated with understanding complex governance proposals and participating in voting processes can lead to decision fatigue and disengagement, contributing to low participation rates (Kahneman, 2011). Cognitive biases such as availability heuristics, where decisions are influenced by easily recalled information (like recent news of exploits or disagreements), and confirmation bias, where individuals favor information confirming their existing beliefs, can skew voting outcomes and hinder constructive deliberation (Kahneman, 2011; Tversky & Kahneman, 1973). These behavioral issues raise ethical concerns about whether decentralized governance truly represents the will of the broader community or is susceptible to manipulation and power concentration, echoing historical difficulties in scaling collective decision-making beyond small groups (Harari, 2014; Zafar, 2025). Case studies like "The DAO" hack highlight the significant risks and vulnerabilities inherent in early decentralized governance experiments (ImmuneBytes, 2023).

Supply Chain Empowerment: Access and Adoption

Blockchain has the potential to empower small-scale producers, particularly in developing nations, by enabling transparent product tracking and potentially ensuring fairer compensation by providing verifiable proof of origin and participation in value chains (Kowalski & Esposito, 2023). This could dismantle existing power imbalances in complex supply chains. However, implementing blockchain-based solutions in these global contexts requires addressing significant challenges related to data standardization, ensuring interoperability between disparate systems, and overcoming barriers related to digital literacy and access to technology among diverse stakeholders (Kowalski & Esposito, 2023).

The perceived complexity and the effort required ("System 2" cost) to adopt and consistently use new blockchain-based tracking and verification systems are significant barriers, particularly for individuals and communities with limited technological infrastructure or experience (Trigsted, 2025). Failure to effectively address these adoption barriers can inadvertently exacerbate existing inequalities, as those with greater resources and technical expertise are better positioned to benefit from the technology. This poses ethical challenges for ensuring inclusive adoption and requires the deliberate creation of accessible and understandable shared "imagined orders" of trust and exchange within these new digital supply chain systems (Carswell, 2024; Harari, 2014; Trigsted, 2025).

Microfinance: Inclusion and Risk

Peer-to-peer microfinance platforms built on blockchain technology could potentially enhance financial inclusion for underserved communities by reducing reliance on traditional intermediaries and lowering transaction costs (Nelson & Charlotte, 2025). This could provide access to financial services for individuals previously excluded from the banking system. However, the implementation of such platforms faces critical hurdles related to scalability, navigating diverse and often complex regulatory compliance requirements across different jurisdictions, ensuring platform security, and guaranteeing accessibility for vulnerable populations (Mhlanga, 2022; Nelson & Charlotte, 2025).

The abstract nature of digital assets and the inherent volatility of cryptocurrencies can be significant deterrents to adoption for individuals with limited financial safety nets, amplifying loss aversion – the tendency to prefer avoiding losses over acquiring equivalent gains (Kahneman, 2011; Nadler et al., 2024). Careful design is needed to prevent algorithmic bias in credit decisions, which could inadvertently discriminate against marginalized groups, and to mitigate security or privacy risks, especially for users who may be more vulnerable to exploitation due to cognitive biases or inexperience with digital financial tools (Nelson & Charlotte, 2025; Zafar, 2025). Furthermore, providing financial tools alone does not solve the multifaceted issues of poverty; socioeconomic factors and behavioral responses necessitate holistic approaches that go beyond technological solutions (Roberts et al., 2021).

Scenario 3: New Income Streams and Economic Models

Blockchain introduces novel ways for individuals to generate income and participate directly in the digital economy, potentially creating new avenues for wealth creation. However, these opportunities are often accompanied by significant volatility, technical barriers, and accessibility issues.

Earning Income: Opportunities and Volatility

Blockchain enables various new forms of income generation, including participating in consensus mechanisms (mining, staking), engaging with decentralized finance (DeFi) through yield farming and lending, creating and trading Non-Fungible Tokens (NFTs), participating in play-to-earn gaming models, monetizing personal data, and engaging in decentralized freelancing and micro-tasking platforms (Borghoff, 2025). These models offer individuals direct participation in digital value creation.

A significant challenge is that these income streams often involve substantial volatility and are susceptible to security risks such as scams and hacks (Borghoff, 2025). Participants are particularly vulnerable to cognitive biases that can influence financial decision-making, including optimism bias (overestimating potential gains), herding behavior (following the actions of a larger group), and framing effects (being influenced by how information is presented) (Kahneman, 2011; Tversky & Kahneman, 1973).

Technical expertise and initial capital requirements for some of these activities can exacerbate the digital divide, raising ethical concerns about fairness and equitable access to these new economic opportunities. The allure of quick or substantial gains ("System 1" thinking) can easily overshadow a more rational assessment of risks and complexities ("System 2" thinking) (Kahneman, 2011). The value of many of these digital assets and income streams is largely based on a shared, often speculative, "imagined" value, making them particularly vulnerable to shifts in collective belief and psychological biases (Carswell, 2024).

Web3 Control and Monetization: Vision and Barriers

The vision of Web3 infrastructure, built upon blockchain and related technologies, is to allow individuals greater control over and the ability to monetize their personal data and digital resources (e.g., selling anonymized appliance data from IoT devices, renting out unused computing power) (Trigsted, 2025). This represents a fundamental shift from the data extraction models prevalent in Web 2.0.

Realizing this vision requires overcoming significant technical hurdles, including the scalability limits of current blockchain networks to handle massive volumes of IoT data on-chain, ensuring robust data privacy and security for interconnected devices, and establishing widely adopted interoperability standards between different platforms (Kowalski & Esposito,

2023; Nadler et al., 2024). The cognitive cost ("System 2" effort) associated with understanding and managing granular data permissions, digital wallets, and resource monetization processes may significantly hinder widespread adoption (Lee et al., 2013; Kahneman, 2011). Furthermore, energy consumption concerns related to underlying blockchain protocols and the technical barriers to entry can also worsen existing inequalities, demanding responsible implementation that carefully considers the cognitive load and accessibility required for this fundamental shift in the "imagined hierarchies" of data control and ownership (Harari, 2014; Kahneman, 2011; Nadler et al., 2024).

Technology Convergence (Blockchain, Web3, IoT): Potential and Complexity

The convergence of blockchain, Web3 principles, and the Internet of Things (IoT) holds immense potential to enable automated and transparent data sharing, peer-to-peer energy trading within smart grids, and efficient resource management facilitated by smart contracts and DAOs (Kowalski & Esposito, 2023; Mhlanga, 2022; Trigsted, 2025). This integration could create highly automated and responsive decentralized systems.

However, this convergence also presents significant technical hurdles, including achieving seamless interoperability between diverse blockchain networks, handling the massive volume and velocity of data generated by IoT devices on-chain or via decentralized storage solutions, and ensuring the robust security of interconnected systems that are vulnerable at multiple points (Kowalski & Esposito, 2023; Nadler et al., 2024). The cognitive complexity ("System 2" overload) of understanding and managing these intertwined systems can be overwhelming for users, potentially deterring adoption and increasing the risk of errors that could lead to unintended consequences, including privacy breaches or the perpetuation of biases embedded within algorithms or data (Iaccarino, 2023; Kahneman, 2011). These integrations raise intricate ethical challenges regarding data privacy, security, and algorithmic bias, demanding careful design of reliable, transparent, and navigable "imagined systems" for human interaction and oversight (Carswell, 2024), Zafar, 2025).

Smart Appliances and IoT Data: Convenience and Risk

The proliferation of internet-connected "smart" appliances and devices in homes and industries generates vast amounts of personal and operational data (Harari, 2014). While offering convenience and efficiency, this data flow presents significant challenges, particularly when considering its integration with decentralized networks.

Protecting this sensitive data on decentralized networks requires robust security measures and clearly defined data ownership and usage protocols (Trigsted, 2025; Zafar, 2025). Users may significantly underestimate the privacy and security risks associated with their smart devices and the data they generate due to familiarity bias (assuming familiar technology is safe) or insufficient "System 2" evaluation of potential threats (Kahneman, 2011; Nelson & Charlotte, 2025; Tversky & Kahneman, 1973). The sensitivity of this data amplifies the risks of privacy violations and the perpetuation of biases if not managed ethically and securely (Nelson & Charlotte, 2025; Zafar, 2025). This necessitates not only technical solutions but also user education designed to overcome cognitive biases and promote a more critical understanding of data privacy in the age of interconnected devices, defining new "imagined contracts" and expectations for our expanding digital selves (Harari, 2014; Kahneman, 2011; Nelson & Charlotte, 2025).

Scenario 4: Navigating Challenges and Ethical Considerations: A Deeper Dive

Realizing blockchain's potential for positive societal impact requires rigorously addressing its inherent challenges and ethical complexities. This necessitates a deep understanding of how the technology interacts with human decisions, societal structures, and our deeply held "imagined orders" (Harari, 2014). The convergence of blockchain with other technologies and its application across various domains surface several critical ethical considerations:

Algorithmic Bias

Bias is a significant concern in any data-driven system, and blockchain applications are not immune. Bias can enter blockchain systems through several pathways: biased training data used for integrated AI algorithms, biased algorithm design choices made by developers, or governance mechanisms (like token-weighted voting) that inherently favor certain groups or concentrate power (Nelson & Charlotte, 2025). This can perpetuate or even amplify real-world inequalities, encoding existing societal "imagined hierarchies" into the digital infrastructure (Harari, 2014; Nelson & Charlotte, 2025). Addressing algorithmic bias in blockchain requires a multi-faceted approach, including rigorous technical audits of algorithms and data, interdisciplinary collaboration involving social scientists and ethicists, and active involvement from diverse community stakeholders in the design and governance processes (Nelson & Charlotte, 2025).

Data Privacy vs. Transparency

The inherent transparency of many public blockchain ledgers, where transaction data is openly viewable, fundamentally clashes with individual privacy rights (Zafar, 2025). While pseudonymity (using a public address instead of real identity) offers a layer of privacy, research has shown that it is often possible to de-anonymize users by analyzing transaction patterns and linking them to external data sources (Chan et al., 2025). Balancing the need for transparency (which fosters trust and auditability) with the fundamental right to privacy is crucial, especially when dealing with sensitive personal or operational data on a blockchain (Zafar, 2025). Privacy-enhancing technologies (PETs) like Zero-Knowledge Proofs (ZKPs) offer potential technical solutions by allowing verification of information without revealing the underlying data, but they often involve trade-offs in terms of cost, computational complexity, and ease of implementation (Kahneman, 2011). Ethical considerations in this domain involve determining who controls privacy levels, who has access to sensitive data, and how to align technological capabilities with evolving societal expectations and legal frameworks regarding data control, challenging our "imagined realities" of data ownership and privacy in a decentralized world (Harari, 2014; Borghoff, 2025).

Security Risks Beyond Cryptography

While blockchain's underlying cryptography provides a strong foundation for security, the broader ecosystem is vulnerable to a range of threats that go beyond cryptographic weaknesses. These include vulnerabilities within smart contracts (as tragically demonstrated by "The DAO" hack) (ImmuneBytes, 2023), network attacks (such as 51% attacks on PoW chains) (OSL, 2025b), economic exploits leveraging vulnerabilities in tokenomics or DeFi protocols, and social engineering tactics targeting individual users to gain access to their digital assets (Tversky & Kahneman, 1973) The immutability of blockchain, while a security feature against tampering, also means that once an error occurs or a system is exploited, it can be extremely difficult or impossible to reverse the transaction or recover lost assets (ImmuneBytes, 2023). This raises complex ethical questions about responsibility, liability, and remediation when security fails within systems often marketed as "trustless," challenging our "imagined security" in a decentralized environment (Harari, 2014; Tversky & Kahneman, 1973).

Environmental Impact

The energy consumption of certain blockchain consensus mechanisms, most notably Proof-of-Work (PoW), raises significant ethical concerns about sustainability and climate impact (Thaler & Sunstein, 2008). The computational race required for PoW consumes vast amounts of electricity, contributing to carbon emissions, particularly when powered by fossil fuels. More energy-efficient alternatives like Proof-of-Stake (PoS) exist and are gaining prominence, but transitions are complex, and other consensus mechanisms have their own distinct trade-offs (Mhlanga, 2022). Evaluating the true environmental impact requires considering not only the energy consumed but also the source of that energy and the full lifecycle analysis of blockchain infrastructure, forcing ongoing debates about the kind of sustainable "imagined future" we are collectively building (Harari, 2014; Thaler & Sunstein, 2008).

Exacerbating Existing Inequalities

Despite its potential for inclusion, blockchain technology can inadvertently exacerbate existing societal inequalities if not developed and deployed thoughtfully. Barriers such as the digital divide (unequal access to internet connectivity, necessary hardware, and digital literacy), wealth concentration (early adopters and those with significant capital can gain disproportionate influence, particularly in token-weighted governance models), and the inherent complexity or poor user experience of many blockchain applications can prevent vulnerable populations from accessing and benefiting from the technology, thereby widening existing socioeconomic gaps (Iaccarino, 2023; Trigsted, 2025). Token-weighted governance, while seemingly democratic, can amplify real-world power imbalances within digital communities, potentially creating new "imagined divisions" based on token holdings (Harari, 2014; Zafar, 2025).

Ethical Frameworks for Evaluation

Analyzing the complex ethical issues arising from blockchain development and deployment requires applying established ethical frameworks to systematically evaluate the moral dimensions of design choices and societal impacts.

Utilitarianism: This framework focuses on weighing the overall benefits and harms of a particular action or technology, aiming to maximize the greatest good for the greatest number. Applying utilitarianism involves assessing the potential positive impacts of blockchain (e.g., financial inclusion, transparency) against the potential negative impacts (e.g., environmental cost, exacerbating inequalities).

Deontology: Deontological ethics emphasizes adherence to moral duties and rights, regardless of the consequences. From this perspective, evaluating blockchain involves considering whether its design and use respect fundamental rights such as privacy, autonomy, and fairness (Zafar, 2025).

Virtue Ethics: This framework focuses on the character and virtues of the actors involved (developers, users, regulators), emphasizing the importance of cultivating virtues like fairness, responsibility, honesty, and justice in the development and use of blockchain technology.

Principlism balances core ethical principles such as autonomy (respecting user choice), beneficence (doing good), non-maleficence (avoiding harm), and justice (ensuring fairness and equitable distribution of benefits and burdens) (Sunstein, 2016). An ethics of care perspective further emphasizes the importance of relationships, empathy, and prioritizing the needs of vulnerable populations in the design and implementation of blockchain solutions. Applying these frameworks helps to move beyond purely technical considerations and systematically evaluate the moral dimensions of blockchain design and deployment in real-world contexts (Sunstein, 2016).

Complex Ethical Dilemmas in Practice

The practical application of blockchain technology often presents complex ethical dilemmas where competing values and principles clash.

Immutability vs. Right to Be Forgotten

The core blockchain principle of immutability, where data is permanently recorded, directly conflicts with evolving privacy regulations like the GDPR, which grant individuals the "right to be forgotten" or the ability to have their personal data erased (Borghoff, 2025). Reconciling the need for permanent, verifiable records with the right to data deletion presents a significant legal and ethical challenge (Borghoff, 2025).

Decentralized Governance Accountability

In decentralized structures like DAOs, determining who is responsible and liable when collective decisions or smart contract vulnerabilities cause harm is a complex ethical and legal question (Zafar, 2025). Decentralization complicates traditional structures of responsibility and accountability that are based on centralized entities (Zafar, 2025).

Censorship Resistance vs. Illicit Use

Balancing the principle of censorship resistance, a key benefit of decentralized networks, with the need to prevent and address illicit activities conducted using blockchain technology (such as money laundering or facilitating illegal markets) is an ongoing ethical tension (Egger et al., 2022).

173

Concrete Recommendations for Responsible Development

Navigating the future of blockchain responsibly requires a multi-faceted approach that prioritizes ethical considerations and inclusive development.

Prioritize Ethical Design: Integrate ethical considerations (including privacy, fairness, security, sustainability, and inclusivity) from the very inception of blockchain projects (often referred to as Ethics-by-Design) (Borghoff, 2025; Zafar, 2025).

Conduct Rigorous Audits: Implement ongoing security audits of code and network protocols, algorithmic bias audits for integrated AI, privacy audits to ensure data protection, and environmental impact assessments (Borghoff, 2025; Nelson & Charlotte, 2025; Thaler & Sunstein, 2008).

Utilize PETs Responsibly: Develop and deploy privacy-enhancing technologies (PETs) like ZKPs to give users meaningful control over their data while mitigating potential for abuse (Kahneman, 2011; Zafar, 2025).

Promote Energy Efficiency: Actively encourage and facilitate the transition to more sustainable consensus mechanisms like Proof-of-Stake (PoS) and invest in research for even more efficient alternatives (Mhlanga, 2022; Thaler & Sunstein, 2008).

Design for Inclusion: Prioritize the creation of user-friendly interfaces, provide accessible educational resources, and explore equitable governance models (such as quadratic voting or delegated systems) that mitigate the influence of wealth and technical expertise (Iaccarino, 2023; Trigsted, 2025; Zafar, 2025).

Foster Interdisciplinary Collaboration: Encourage collaboration among technologists, ethicists, social scientists, legal experts, policymakers, and diverse community stakeholders throughout the development and deployment lifecycle (Borghoff, 2025; Nelson & Charlotte, 2025; Sunstein, 2016).

Establish Clear Governance and Accountability: For blockchain applications with real-world impact, explore hybrid governance models or legal wrappers for DAOs that establish clear lines of responsibility and accountability (Zafar, 2025).

Advocate for Thoughtful Regulation: Engage with policymakers to develop nuanced, adaptive regulations that foster innovation while effectively protecting users, addressing ethical concerns, and providing legal certainty (Borghoff, 2025).

Educate Users Critically: Empower users with knowledge about both the potential benefits and the risks and ethical considerations of blockchain technology; promote critical engagement and informed decision-making (Kahneman, 2011; Nelson & Charlotte, 2025; Tversky & Kahneman, 1973).

Collaboration, Pragmatism, and Human-Centricity

Shaping blockchain's future demands ongoing collaboration, critical thinking, and a steadfast commitment to a human-centered approach. Acknowledging the technology's current limitations (including scalability challenges, security vulnerabilities, and regulatory uncertainties) alongside its potential benefits is crucial for pragmatic development (Borghoff, 2025; Nadler et al., 2024; Zafar, 2025). Integrating insights from behavioral science, ethics, and diverse disciplines allows us to harness blockchain's transformative power responsibly, ensuring that technological advancements are aligned with human values and societal well-being (Kahneman, 2011; Sunstein, 2016; Zafar, 2025). The ultimate goal is not merely technological advancement for its own sake, but the deliberate construction of more just, equitable, and sustainable digital systems that are cognizant of the profound influence of our shared "imagined orders" (Harari, 2014). This requires continuous dialogue, active and inclusive participation from all stakeholders, and a collective commitment to designing and implementing technology that genuinely serves humanity, grounded in a realistic understanding of both its immense potential and its inherent challenges.

Chapter 7: Navigating the Blockchain Information Landscape

In the dynamic and rapidly evolving world of blockchain technology, accessing credible and reliable information is paramount. The landscape is vast, often filled with technical jargon, speculative claims, and diverse perspectives. To truly understand blockchain, its potential, and its limitations, one must learn to effectively navigate this information environment, discerning reliable sources from noise (Barthel, 2020; Kiely & Robertson, 2020). This chapter serves as a guide to identifying established news outlets, research platforms, and community resources, while also providing essential techniques for evaluating the information encountered to ensure accuracy and build a well-rounded understanding.

Identifying Credible Information Sources

Navigating the blockchain information landscape begins with identifying reputable sources that offer accurate reporting, in-depth analysis, and verified data. While the space is constantly changing, several platforms have established themselves as valuable resources for different levels of engagement and technical understanding.

For General Information & Analysis

These sources are often the first point of contact for individuals seeking to understand current events, market trends, and foundational concepts within the blockchain space. They typically provide news updates, introductory explanations, and high-level analysis.

CoinDesk

Often referred to as a leading source for cryptocurrency and blockchain news, CoinDesk offers extensive reporting, market data, and analysis covering a wide range of topics within the industry (Coindesk, n.d) They are recognized for their journalistic approach to covering the space.

The Block

Known for its focus on data-driven research and analytics, The Block provides a more technical and analytical perspective. Their reports and data dashboards are valuable for those seeking deeper insights into market structure, funding trends, and protocol activity (The Block, n.d).

Decrypt

Decrypt aims to make blockchain and cryptocurrency concepts accessible to a broader audience. They offer clear explanations, engaging content, and focus on the cultural and societal impacts of decentralized technologies, making it a good starting point for beginners (Decrypt, n.d.).

For a Deeper Dive

For those looking to move beyond general news and engage with more detailed analysis, expert opinions, and technical discussions, these platforms offer more in-depth content.

Cointelegraph

This platform provides comprehensive coverage of the crypto world, including breaking news, expert commentary, technical analysis, and educational resources. They cover a wide range of projects and developments within the ecosystem (Cointelegraph, n.d.).

BeInCrypto

BeInCrypto offers news, reviews, and guides on various aspects of the cryptocurrency and blockchain space. They also provide educational content and resources for those looking to deepen their understanding (BeInCrypto, n.d.).

Beyond News Outlets: Research and Community

Beyond journalistic sources, academic research and community platforms play crucial roles in the blockchain information ecosystem, offering peer-reviewed studies, technical discussions, and diverse perspectives.

Academic Journals

For rigorous, peer-reviewed research, academic journals are invaluable. Publications like the International *Journal of Blockchain Research* and *Ledger* offer in-depth studies on the technical underpinnings, economic implications, and societal impacts of blockchain technology, providing evidence-based insights (Ledger, n.d.).

Industry Reports

Reports published by reputable consulting firms and industry leaders such as Deloitte, PwC, and Gartner provide valuable insights into blockchain adoption trends, industry-specific applications, challenges, and market forecasts (Banerjee et al., 2015; Gartner, n.d.; Grewal-Carr et al., n.d.; Zafar, 2025). These reports often synthesize information from various sources and offer a broader view of the industry landscape.

Community Forums and Subreddits

Online communities like r/Blockchain and r/ethdev on platforms such as Reddit can be excellent resources for real-time discussions, asking questions, sharing resources, and gaining insights from active participants in the space (Reddit, n.d.). However, it is crucial to approach information from these sources with critical caution, as it represents individual opinions and is not subject to formal verification processes (Bank for International Settlements, n.d.).

Fact-Checking and Evaluating Information

In an information environment as dynamic and sometimes speculative as blockchain, developing strong fact-checking and evaluation skills is essential. Not all sources are equally reliable, and misinformation can spread quickly. Beyond simply considering the source, a more detailed approach to verifying information is necessary.

Tips for Evaluating Information

Consider the Source and Its Reputation: As a foundational step, evaluate the source's track record. Is it an established publication with a history of accurate reporting? Does it have editorial standards and fact-checking processes? Be wary of anonymous sources or platforms that lack transparency about their contributors or funding (Baker & Wurgler, 2007).

Check for Author Expertise and Affiliations: Investigate the author's credentials and experience in the blockchain field. Do they have relevant technical knowledge, academic background, or industry experience? Be mindful of potential conflicts of interest based on their affiliations (Baker & Wurgler, 2007).

Look for Evidence and Supporting Data: Reliable claims are typically backed by evidence. Does the information presented include links to data, research papers, code repositories, or other verifiable sources? Be skeptical of claims that rely solely on anecdotal evidence (Baker & Wurgler, 2007).

Cross-Reference Information: Do not rely on a single source for critical information. Cross-reference the information with multiple reputable and independent sources. If a claim is significant, it is likely being reported and discussed by several credible outlets (Iaccarino, 2023).

Be Wary of Hype, Unrealistic Promises, and Emotional Language: The blockchain space can be prone to hype and speculation. Be cautious of sources that make exaggerated claims about potential profits, promise unrealistic returns, or use highly emotional or promotional language. These can be indicators of bias or attempts to mislead (Baker & Wurgler, 2007).

Identify Potential Biases: Every source may have biases, whether intentional or unintentional. Consider the source's business model (e.g., is it funded by specific projects?), its target audience, and its historical stance on certain topics. Industry reports, while valuable, may be influenced by the interests of the companies publishing them or the clients they serve (Baker & Wurgler, 2007). Academic research, while peer-reviewed, can sometimes have limitations in scope or be influenced by funding sources. Community forums, while offering diverse views, can be susceptible to groupthink or manipulation by motivated actors (Bank for International Settlements, n.d.).

Check Dates and Timeliness: The blockchain space moves rapidly. Ensure the information you are accessing is current and relevant. Outdated information may no longer be accurate due to technological advancements or market changes (Baker & Wurgler, 2007).

Understand the Difference Between Fact, Opinion, and Analysis: Clearly distinguish between factual reporting, expert opinions, and analytical pieces. While opinions and analysis can provide valuable perspective, they should not be treated as verified facts (Baker & Wurgler, 2007).

Detecting Misinformation

Actively detecting misinformation requires a proactive approach and a skeptical mindset.

Reverse Image Searches: If an article includes images or graphics that seem questionable, perform a reverse image search to see if they have been used out of context or are associated with different information (Iaccarino, 2023).

Fact-Checking Websites: Utilize established fact-checking websites, although dedicated blockchain fact-checkers are still developing. Apply general fact-checking principles to claims made about blockchain (Iaccarino, 2023).

Trace Claims Back to Original Source: If a source refers to a study, report, or statement, try to find the original source to verify that the information is being accurately represented (Iaccarino, 2023).

Look for Red Flags: Be alert for poor grammar, spelling errors, unprofessional website design, or aggressive advertising, which can sometimes indicate less reputable sources (Baker & Wurgler, 2007).

By employing these fact-checking techniques and maintaining a critical perspective, individuals can navigate the blockchain information landscape more effectively and build a more accurate understanding of this complex technology.

Understanding Blockchain Performance

For developers, investors, and anyone interested in the practical viability of blockchain technology, understanding its performance characteristics is crucial. Performance metrics determine how efficiently a blockchain network can process transactions and scale to meet demand.

Credible Reports and Resources on Blockchain Performance

Evaluating blockchain performance requires consulting technical reports, academic studies, and industry analyses that delve into key metrics and comparative analyses.

Hyperledger Fabric Performance Metrics White Paper: Published by the Linux Foundation, this paper provides a foundational framework and defines key metrics for measuring the performance of blockchain networks, particularly in enterprise contexts. It outlines methodologies for conducting performance evaluations (Hyperledger Performance and Scale Working Group, n.d.).

Performance Analysis of Blockchain Platforms: Academic papers, such as those available through university research repositories, often offer comparative analyses of different blockchain platforms, evaluating their performance based on various metrics under controlled conditions (Gubbi et al., 2013).

Blockchain Performance Issues and Limitations: Articles and reports from blockchain security and development companies often explore the practical challenges and limitations in accurately measuring blockchain performance in real-world scenarios and discuss factors that can impact it (Hovard, 2023).

BEDAO: A White Paper for Managing Blockchain Communities: While primarily focused on DAO governance, this white paper touches upon the underlying performance considerations necessary for effective and scalable decentralized community management (Carswell, 2024).

Articles and Blogs (Focused on Performance)

Industry blogs and articles from reputable blockchain companies and research groups often discuss the importance of performance, benchmarking, and scalability solutions.

Unlocking Blockchain's Full Potential: The Critical Role of Performance Benchmarking: Articles like this emphasize why systematically measuring and comparing blockchain performance is essential for understanding its capabilities and limitations for different use cases (Savage, 2016).

Why Blockchain Performance Matters: Discussions from industry consortia highlight the critical link between blockchain performance, its potential for mainstream adoption, and the ongoing need for developing effective scalability solutions (Atzei et al., 2015).

Resources for Blockchain Performance Analysis

Tools

Specialized tools are available for conducting performance testing and analysis on blockchain networks.

Caliper

An open-source blockchain performance benchmarking framework that allows users to measure the performance of different blockchain platforms under various workloads and configurations (Stoll et al., 2017).

Hyperledger Cello

A tool for deploying and managing blockchain networks, which can be used to set up environments for performance testing and analysis (Gillis, 2023; Hyperledger, n.d.).

Key Factors to Consider When Evaluating Blockchain Performance

When evaluating the performance of a blockchain network, several key factors are typically considered:

Transaction Throughput: This is a fundamental metric, representing the number of transactions the blockchain can process per second (TPS). Higher throughput is generally required for applications with high transaction volumes (Karim et al., 2025).

Latency: This refers to the time it takes for a transaction to be confirmed and added to the blockchain. Lower latency is important for applications requiring fast finality (Karim et al., 2025).

Scalability: This is the ability of the blockchain to handle increasing transaction volumes and network load without a significant degradation in performance. Scalability solutions, such as Layer 2 protocols, are designed to improve this (Atzei et al., 2015).

Resource Consumption: This includes the amount of computing power, memory, and storage required to run a node on the blockchain network. High resource consumption can impact decentralization by making it more difficult for individuals to run full nodes (Karim et al., 2025).

Consensus Mechanism: The method used to validate transactions and add blocks to the blockchain (e.g., Proof-of-Work, Proof-of-Stake) significantly impacts performance, energy consumption, and scalability trade-offs (Conway, 2022; Hovard, 2023).

By exploring these resources, utilizing performance analysis tools, and considering these key factors, one can gain a deeper understanding of blockchain performance and its implications for various use cases and the future development of the technology.

Navigating the blockchain information landscape effectively requires a combination of identifying credible sources, employing rigorous fact-checking techniques, and understanding the technical nuances of blockchain performance. By critically evaluating information from diverse sources, being mindful of potential biases, and utilizing available tools and resources, individuals can build a more accurate and comprehensive understanding of blockchain technology. This critical engagement is essential for making informed decisions, whether as a developer, investor, or simply an interested observer in this rapidly evolving field.

Chapter 8: Glossary of Key Terms

AI-Generated Content: Content, such as text, images, or music, created by AI algorithms.

AI Ethics: The study of the moral issues that arise from the development and use of AI.

AI Model Marketplace: A platform where AI models and related resources like datasets can be shared, bought, and sold.

Algorithmic Bias: Systematic errors in AI or algorithmic outputs that result in unfair outcomes, often reflecting biases in the training data or design.

Anchoring Bias: A cognitive bias that describes our tendency to over-rely on the first piece of information we receive (the "anchor") when making decisions, even if that information is not relevant or reliable.

Artificial Intelligence (AI): The simulation of human intelligence processes by machines, especially computer systems.

Auditability: The ability to track and verify transactions and data changes.

Availability Bias: The availability bias, also known as the availability heuristic, is a mental shortcut that occurs when we overestimate the likelihood of events that are more easily recalled, often because they are vivid, recent, or emotionally impactful.

Availability Heuristic: A mental shortcut where people estimate the probability of an event based on how easily examples come to mind.

Behavioral Economics: A field that combines insights from psychology and economics to understand how cognitive biases, heuristics, and emotions influence economic decisions.

Behavioral Science: The study of human behavior, often drawing on psychology, economics, and neuroscience.

Bias: Generally, refers to a tendency, inclination, or prejudice toward or against something or someone. It's often seen as unfair and can be based on stereotypes rather than objective information. Biases can be conscious (explicit) or unconscious (implicit).

Bias in AI: When AI algorithms produce outcomes that are unfairly prejudiced towards particular groups.

Blockchain Performance: How efficiently a blockchain network can process transactions and scale.

Blockchain Psychology: Blockchain psychology is the study of how human behavior interacts with and influences the adoption, development, and use of blockchain technology. It delves into the psychological factors that drive user engagement, trust, and decision-making within blockchain-based systems.

Blockchain Technology: A secure, decentralized software system used for storing and managing data to securely record transactions.

Byzantine Fault Tolerance (BFT): A concept in distributed computing that describes a system's ability to function correctly even if some components fail or act maliciously.

Carbon Footprint Tracking: The process of measuring and monitoring the total greenhouse gas emissions caused directly or indirectly by a person, organization, event, or product.

Cognition: Cognition refers to the mental processes involved in acquiring knowledge and understanding. It encompasses a wide range of higher-level brain functions that allow us to interact with the world.

Cognitive Alignment: Designing blockchain solutions to minimize cognitive load and align with human cognitive processes.

Cognitive Bias: Systematic deviation from normal or rational judgment. These are mental shortcuts that can lead to systematic errors in judgment, influencing our actions and choices in ways we may not even realize.

Cognitive Dissonance: Mental discomfort, stress, unease, tension, or guilt, that occurs when a person's behavior contradicts their beliefs.

Cognitive Load: The total amount of mental effort being used in the working memory.

Cognitive Psychology: The scientific study of mental processes such as attention, language use, memory, perception, problem solving, creativity, and thinking.

Confirmation Bias: Confirmation bias is a cognitive bias that de- scribes our tendency to favor information that confirms our existing beliefs and to ignore or downplay information that contradicts them. It's a way of selectively filtering information to reinforce what we already think is true.

Consensus Mechanism: Automated protocols or algorithms that enable distributed nodes in a network to collectively agree on the correct state of the blockchain and the legitimacy of transactions.

Consortium Blockchain: A permissioned blockchain network governed by a group of pre-selected organizations or institutions.

Cryptographic Security: The use of advanced mathematical techniques to secure transactions and data.

Cultural Intelligence: The ability to understand and effectively interact with people from different cultures.

Cultural Context: The influence of cultural values, norms, and existing social structures on behavior and technology adoption.

DAO (Decentralized Autonomous Organization): An organization run through rules encoded as computer programs called smart contracts.

Data Integrity: The accuracy, completeness, and consistency of data.

Data Provenance: The history of the data, including where it came from and what transformations it has undergone.

Decentralization: Power and control are distributed among several nodes or participants rather than a central authority or agency.

Decentralized Autonomous Organization: An autonomous blockchain function that follows rules and regulations to ensure transparency and prevent participant influence.

Decentralized Applications (dApps): Applications that run on a decentralized block- chain network, rather than a centralized server.

Decentralized Finance (DeFi): A financial system built on blockchain technology that aims to recreate traditional financial instruments in a decentralized architecture, outside of the control of central banks and traditional financial institutions.

Decentralized Governance: Decentralized governance means people make decisions and run a blockchain network instead of a central authority. This architecture ensures that all network participants have modification authority.

Decentralized Identity (DID): A type of identifier that is globally unique, cryptographically verifiable, and designed to be controlled by the individual or entity that owns it, independent of centralized authorities.

Decentralized Marketplace: An online platform for buying and selling where transactions occur directly between participants without a central authority.

Decentralized Stablecoin: A type of cryptocurrency designed to maintain a stable value (typically pegged to a fiat currency like the US dollar or another asset) without relying on a central issuer, custodian, or traditional financial reserves managed by a single entity. They use automated systems built on blockchain technology.

Decision Hygiene: Strategies for individuals and organizations to reduce noise's impact on decision-making for educated, cost-effective decision-making.

Deep Learning (DL): A subset of machine learning that uses artificial neural networks with multiple layers to analyze various factors of data.

DeFi (Decentralized Finance): Financial services offered on blockchain networks, without traditional financial intermediaries.

Digital Divide: The gap between demographics and regions that have access to modern information and communications technology, and those that don't or have restricted access.

Digital Likeness: A digital representation of a person's appearance or voice.

Digital Literacy: The ability to use information and communication technologies to find, evaluate, create, and communicate information, requiring both cognitive and technical skills.

Digital Rights Management (DRM): Technologies used to control the use, modification, and distribution of copyrighted works.

Distributed Ledger Technology (DLT): A digital system for recording the transaction of assets in which the transactions and their details are recorded in multiple places at the same time (Di Ciccio, 2024).

dApp (Decentralized Application): An application that runs on a decentralized network, typically a blockchain, without central control.

Echo Chamber: A closed information environment where people are primarily exposed to information and perspectives that reinforce their existing beliefs, leading to the exclusion of alternative viewpoints and a distorted perception of reality.

Emotional Intelligence: The capacity to recognize, understand, and manage one's own emotions and those of others.

Emotional Resonance: Designing blockchain solutions to evoke positive emotions and mitigate negative ones in users.

Endowment Effect: A cognitive bias that causes individuals to value an owned object higher than its market value, often irrationally. Essentially, people place higher value on something simply because they own it.

Enhanced AI Transparency and Explainability: The ability to understand and interpret how an AI system arrived at a particular decision.

Ethics-by-Design: An approach to technology development that integrates ethical considerations throughout the design process.

Excessive Coherence: Also known as the coherence bias, is a cognitive bias that describes our tendency to prefer narratives and explanations that are simple, consistent, and internally coherent, even if they are incomplete or inaccurate. We tend to favor stories that "make sense" over those that are complex or contradictory, even if the latter are closer to the truth.

Explainability in AI: The ability to understand and interpret how an AI system arrived at a particular decision.

Fact-Checking: The process of verifying the accuracy of information.

FOMO (Fear Of Missing Out): Anxiety that an exciting or interesting event may currently be happening elsewhere, often fueled by availability bias in volatile markets.

Framing Effect: A cognitive bias where people react to choices differently depending on how they are presented or framed, even if the options are objectively the same. The way information is presented, or the "frame" it's put in, can significantly influence our choices and judgments.

Governance Tokens: A specific type of cryptocurrency that grants their holders voting rights and influence over the development and decision-making processes of a particular blockchain project, protocol, or a Decentralized Autonomous Organization (DAO).

Greenwashing: Making a product, service, or company appear more environmentally friendly than it really is.

Group Think: A psychological phenomenon that occurs within a group when the desire for harmony or consensus overrides a realistic appraisal of alternatives. In essence, members prioritize group cohesion and agreement over critical thinking and individual judgment.

Hashing: The process of converting data of any size into a fixed-size string of bytes, serving as a unique digital fingerprint.

Hashcash: A Proof-of-Work pioneer developed by Adam Back to deter email spam.

Heuristic: A mental shortcut or rule of thumb that simplifies decision-making, especially in complex or uncertain situations. Heuristics allow us to make judgments quickly and efficiently, but they can also lead to cognitive biases and errors in judgment.

Heuristics: Mental shortcuts or rules of thumb used to make decisions quickly and efficiently.

HODLing: A term in cryptocurrency referring to holding onto assets during market volatility, often influenced by loss aversion.

Human Algorithm: A term used in the text to represent the complex interplay of human behavior, cognition, and social dynamics underlying decentralized systems.

Human-Centered Design: An approach to design that focuses on understanding and addressing human needs and capabilities.

Hype Cycle: A graphical representation of the maturity, adoption, and social application of specific technologies.

Imagined Orders: Shared beliefs and myths that enable large groups of humans to cooperate effectively (term coined by Yuval Noah Harari).

Immutability: Once data (like transactions) has been recorded onto the blockchain ledger and verified, it cannot be altered, changed, deleted, or tampered with.

Impermanent Loss: A potential, temporary loss of funds experienced by liquidity providers in decentralized finance due to volatility in the trading pair.

In Real Life (IRL): This refers to experiences whereby individuals are in a physical location, often with other individuals.

Industry Report: A publication by a consulting firm or industry group analyzing trends, market size, and outlook for a specific sector.

Intermediaries: Middlemen, or third-party providers, often eliminated or reduced in the role of blockchain transactions processing.

Interoperability: The ability of different blockchain networks to communicate, interact, and share data or value (like tokens) with each other seamlessly.

Interoperability (Blockchain): The ability of different blockchain networks to exchange information and value.

Intellectual Property (IP): Creations of the mind, such as inventions, literary and artistic works, designs, and symbols, names, and images used in commerce.

IoT (Internet of Things): The network of physical objects—"things"—that are embedded with sensors, software, and other technologies for the purpose of connecting and exchanging data with other devices and systems over the internet.

Latency: The time delay between initiating a transaction and its confirmation on the blockchain.

Layer 1 Scaling Solutions: Strategies that improve the processing capabilities directly on the main blockchain.

Layer 2 Scaling Solutions: Strategies that move transaction processing off-chain, using the main blockchain for final settlement and security.

Ledger: A digital system for recording the transaction of assets in which the transactions and their details are recorded in multiple places at the same time.

Loss Aversion: A cognitive bias that describes how people tend to feel the pain of a loss more strongly than the pleasure of an equivalent gain. Essentially, the negative feeling of losing $100 is greater than the positive feeling of gaining $100.

Machine Learning (ML): A type of AI that allows computer systems to learn from data and improve without being explicitly programmed.

Mental Accounting: Developed by Nobel laureate Richard Thaler, it describes the tendency for people to treat money differently depending on subjective factors like its origin (e.g., salary, gift, bonus) or its intended use (e.g., rent money, vacation fund), rather than treating all money as interchangeable (fungible).

Merkle Tree: A tree-like data structure used for efficient verification of large amounts of data, often used in blockchains.

Misinformation: False or inaccurate information, especially that which is deliberately intended to deceive.

Natural Language Processing (NLP): A branch of AI that helps computers understand, interpret, and manipulate human language.

Neural Network: A series of algorithms that endeavors to recognize underlying relationships in a set of data through a process that mimics the way the human brain operates.

Node: A computer connected to the blockchain network that holds a copy of the distributed ledger. They are fundamental components that run the blockchain's software protocol and collectively form the network's infrastructure. By distributing these tasks across many nodes, the blockchain achieves decentralization, security, and resilience, as there is no single point of failure or control.

Noise: Unexpected, random judgment changes. Noise is an outlier in aggregated decisions that often goes undetected.

Non-Fungible Tokens (NFT): Non-fungible tokens (NFTs) are unique digital assets with distinct values.

Nudge: A concept in behavioral economics, political theory, and behavioral design which proposes indirect suggestions to try to influence the motives, incentives, and decision making of groups and individuals.

Off-Chain: Any transaction, data storage, or computation that takes place outside of the main blockchain network (Layer 1).

On-Chain: Transactions that are directly recorded, validated, and stored on the main blockchain itself, benefiting from its full security and transparency but often limited by its speed and cost.

Peer-Reviewed Research: Academic work that has been evaluated by experts in the same field before publication.

PETs (Privacy-Enhancing Technologies): Technologies that incorporate privacy and data protection into the design of information systems and services.

Predictive Analytics: The use of data, statistical algorithms, and machine learning techniques to identify the likelihood of future outcomes based on historical data.

Private Blockchain: A permissioned blockchain network where access and participation are restricted and controlled by a central authority or organization.

Proof-of-Authority (PoA): A consensus mechanism that relies on pre-selected, reputable validators.

Proof-of-Stake (PoS): A consensus mechanism where participants (validators) stake a certain amount of cryptocurrency as collateral to participate in the consensus process and validate transactions.

Proof-of-Work (PoW): A consensus mechanism where participants (miners) solve complex cryptographic puzzles to validate transactions and add new blocks to the blockchain.

Prospect Theory: Developed by psychologists Daniel Kahneman and Amos Tversky, it describes how people make choices between probabilistic alternatives involving risk (potential gains and losses).

Provenance (Data): The history of data, including its origin and any transformations.

Provenance (Digital): The history of ownership and location of a digital asset.

Prosumer: An individual who both produces and consumes a good or service, such as energy in a microgrid.

Pseudonymity: Using a pseudonym (like a wallet address) rather than real-world identity.

Public Blockchain: A permissionless blockchain network where anyone can participate, read the ledger, submit transactions, and participate in the consensus process.

Rational Actor Model: A model in traditional economics that assumes individuals make decisions based on perfect information to maximize their utility.

Regulatory Landscape: The body of laws, rules, and guidelines that govern blockchain technology and crypto assets within a specific jurisdiction.

Renewable Energy Trading: The buying and selling of energy generated from renewable sources, often facilitated by blockchain.

Representativeness Heuristic: A mental shortcut people use when making judgments about the probability of an event that is based on how similar the event or object is to a prototype or stereotype they already have in mind.

Resource Consumption (Blockchain): The computing power, memory, and storage needed to run a blockchain node.

Rollups: A Layer 2 scaling technique that bundles multiple off-chain transactions into a single transaction and posts a compressed summary or proof onto the main chain.

Satoshi Nakamoto: The pseudonymous creator(s) of Bitcoin.

Scalability: The ability of a blockchain network to handle a growing number of transactions efficiently without compromising speed or increasing costs.

Scalability (Blockchain): The ability of a blockchain to handle increasing transaction volume and users.

Self-Sovereign Identity (SSI): A model of digital identity where individuals have complete control over their personal data and how it is shared.

Sharding: A Layer 1 scaling technique that divides the network state and processing load into smaller partitions (shards) to enable parallel processing.

Smart Contract: A blockchain-based program or mechanism that automatically executes a contract under predetermined conditions. Also called self-executing contracts.

Social Intelligence: The ability to navigate and understand social interactions and dynamics.

Social Science: The scientific study of human society and social relationships.

Stablecoin: A type of cryptocurrency specifically designed to maintain a stable market value. Unlike highly volatile cryptocurrencies like Bitcoin or Ethereum, stablecoins attempt to peg their value to an external reference point, most commonly to fiat(currency), another cryptocurrency or hard assets like gold.

State Channels: A Layer 2 scaling technique that allows direct off-chain transactions between participants, with only the opening and closing of the channel settled on the main chain.

Stereotype: A fixed, generalized, often oversimplified, and potentially biased belief or assumption about the characteristics attributed to individuals based solely on their membership in a particular group (e.g., based on race, gender, age, occupation, etc.).

Supply Chain Transparency: The degree to which information about a supply chain is available to stakeholders.

Synthetic Media: Media that is artificially generated or manipulated, such as deepfakes.

System 1 Thinking: Fast, intuitive, and emotional thinking (from Daniel Kahneman's "Thinking, Fast and Slow").

System 2 Thinking: Described by Daniel Kahneman as the slow, effortful, infrequent, logical, calculating, and conscious mode of thought. System 2 monitors the output of System 1 and takes over when tasks become too difficult for System 1.

System Noise: The inherent variability and inconsistencies in user behavior, interactions, and the spread of information that emerge from the platform's architecture, features, and the collective dynamics of its users.

The Theory of Blockchain Psychology: Explains the intricate relationship between blockchain technology and human behavior, recognizing that technology's true potential lies in its ability to align with and enhance human experiences.

Throughput: The rate at which a system can process transactions (e.g., transactions per second).

Throughput (Blockchain): The number of transactions a blockchain can process per second (TPS).

Token: A digital asset that represents a unit of value or utility on a blockchain.

Token-Weighted Voting: A governance system where voting power is based on token holdings.

Tokenized Incentives: The use of blockchain-based tokens to reward positive contributions or desired behaviors within a network or platform.

Transparency: The characteristic of most public blockchains where all participants can view the transaction history of the ledger.

User-Centric Design: An approach to design that focuses on understanding and addressing human needs and capabilities.

Utility: Refers to the benefits a user (NFT, crypto owner) receives beyond digital asset ownership. These advantages vary greatly depending on the project and its goal.

Vitalik Buterin: Co-founder of Ethereum.

Web 1.0: In the mid-1990s, the World Wide Web began as Web 1.0. It led to the internet we know today but operated differently. Websites were mostly static, one-way websites.

Web 2.0: User-generated content, interactivity, and social interaction define the internet today. Compared to Web 1.0, it revolutionized internet use in the early 2000s.

Web 3.0: Often referred to as the decentralized web, envisioning an internet where users have greater control over their data, identity, and online interactions, often built on blockchain technology.

Web3: Next-generation internet idea Web3 (Web 3.0) is emerging. Decentralized, open, and user-centric, it promises to change the Web 2.0 landscape currently controlled by IT giants.

White Paper: An informative document issued by a company or non-profit to promote or highlight the features of a solution, product, or service.

Zero-Knowledge Proof (ZKP): A cryptographic method by which one party (the prover) can prove to another party (the verifier) that a given statement is true, without revealing any information beyond the validity of the statement itself.

References

Ahmed, D. (2023). *Blockchain cryptographic security, hashing and digital signature*. eGov Standards. https://egovstandards.gov.in/sites/default/files/2023-05/Blockchain%20Cryptographic%20Security%2C%20Hashing%20and%20Digital%20Signature.pdf

Ahmed, S. (2023). *Study of cryptographic techniques adopted in blockchain*. ResearchGate. https://www.researchgate.net/publication/372120695

Ajzen, I. (1991). The theory of planned behavior. *Organizational Behavior and Human Decision Processes*, *50*(2), 179-211. https://www.sciencedirect.com/science/article/abs/pii/074959789190020T

Akter, A., Arobee, A., Al Adnan, A., Auyon, O., Islam, A. A., & Akter, F. (2025). *Blockchain as a platform for artificial intelligence (AI) transparency*. arXiv. https://arxiv.org/abs/2503.08699

Aldoubaee, A., Hassan, N. H., & Rahim, F. A. (2024). *A systematic review on blockchain scalability*. ResearchGate. https://www.researchgate.net/publication/374496563

American Psychological Association. (2021, June 30). *What use is technology if no one uses it? The psychological factors that influence technology adoption decisions in oil and gas*. Technology, Mind, and Behavior, 2(1). https://doi.org/10.1037/tmb0000027

Ammar, Z. (2025). Reconciling blockchain technology and data protection laws: regulatory challenges, technical solutions, and practical pathways. *Journal of Cybersecurity, 11*(1). https://academic.oup.com/cybersecurity/article/doi/10. 1093/cybsec/tyaf002/8024082

Asch, S. E. (1956). Studies of independence and conformity: A minority of one against a unanimous majority. *Psychological Monographs: General and Applied, 70*(9), 1–70. https://www.researchgate.net/publication/254732835

Back, A. (2002). *Hashcash - A denial of service counter-measure.* http://www.hashcash.org/papers/hashcash.pdf

Bandura, A. (1977). Self-efficacy: Toward a unifying theory of behavioral change. *Psychological Review, 84*(2), 191–215. https://www.sciencedirect.com/science/article/abs/pii/ 0146640278900024

Banerjee, A. V., Duflo, E., Glennerster, R., & Kinnan, C. (2015). The miracle of microfinance? Evidence from a randomized evaluation. *American Economic Journal: Applied Economics, 7*(1), 241-273. https://www.aeaweb.org/articles?id=10.1257/app.201 30533

Bank for International Settlements. (n.d.). *The crypto ecosystem: key elements and risks.* https://www.bis.org/publ/othp72.pdf

Barik, K., Misra, S., Ray, A. K., & Sukla, A. (2023). A blockchain-based evaluation approach to analyse customer satisfaction using AI techniques. *Heliyon*, *9*(6), e16766. https://pmc.ncbi.nlm.nih.gov/articles/PMC10245048/

Barthel, M. (2020). *Many Americans believe fake news is sowing confusion*. Pew Research Center's Journalism Project. https://www.pewresearch.org/journalism/2016/12/15/many-americans-believe-fake-news-is-sowing-confusion/

Bialas, S. (2024). *Blockchain use cases: Top applications across industries - Ulam Labs*. Ulam Labs. https://www.ulam.io/blog/exploring-blockchain-use-cases-across-industries

Borghoff, U., Bottoni, P., & Pareschi, R. (2025). *Human-artificial interaction in the age of agentic AI: A system-theoretical approach*. arXiv. https://arxiv.org/pdf/2502.14000

Brookings Institution. (2025, April 10). *The hidden danger of re-centralization in blockchain platforms*. https://www.brookings.edu/articles/the-hidden-danger-of-re-centralization-in-blockchain-platforms/

Buterin, V. (2014). *Ethereum white paper: A next-generation blockchain and decentralized application platform*. Ethereum. https://ethereum.org/en/whitepaper/

Carswell, C. J. (2024). *BE DAO: A whitepaper for managing blockchain communities: Integrating behavioral economics, decision hygiene, and global ethics.* BlocPsych.

Chan, W., Gai, K., Yu, J., & Zhu, L. (2025). Blockchain-assisted self-sovereign identities on education: A survey. *Blockchains, 3*(1), 3. https://doi.org/10.3390/blockchains3010003

Chen, B., Ma, L., Xu, H., Ma, J., Hu, D., Liu, X., Wu, J., Wang, J., & Li, K. (n.d.). *A comprehensive survey of blockchain scalability: Shaping inner-chain and inter-chain perspectives.* arXiv.org e-Print archive. https://arxiv.org/html/2409.02968v1

Conway, L. (2022). *Proof-of-work vs. proof-of-stake: Which is better?* Blockworks.co. https://blockworks.co/news/proof-of-work-vs-proof-of-stake-whats-the-difference

Debutinfotech. (2025). *A comprehensive guide to different types of DAOs.* https://www.debutinfotech.com/blog/different-types-of-daos

DeJeu, E. (2025). *New Tepper research highlights the power of distributed ledger technology (DLT)...* Center for Intelligent Business. https://www.cmu.edu/intelligentbusiness/news-events/distributed-ledgers-research.html

Di Ciccio, C. (2024). *Blockchain and distributed ledger technologies.* ResearchGate.

https://www.researchgate.net/publication/377843175_
Blockchain_and_Distributed_Ledger_Technologies

Dixit, S., & Jagid, J. (2025). *Exploring smart contracts and artificial intelligence in FinTech.* ResearchGate.
https://www.researchgate.net/publication/389518214

Egger, R., Neuburger, L., & Mattuzzi, M. (2022). Data science and ethical issues: Between knowledge gain and ethical responsibility. In *Data science and ethical issues* (pp. 57-78). Springer.
https://www.researchgate.net/publication/358255149

Fan, J. (2024). *Issues and reflections on DAO: Governance challenges and solutions - AIFT.* AIFT.
https://hkaift.com/issues-and-reflections-on-dao-governance-challenges-and-solutions/

Finance Technology Insights. (2024). *The role of AI in improving blockchain scalability and efficiency.*
https://globalfintechseries.com/featured/role-of-ai-in-optimizing-blockchain-scalability/

Flatworld Solutions. (2025). *Big data and blockchain analytics - Flatworld solutions.*
https://www.flatworldsolutions.com/data-science/articles/big-data-blockchain-analytics-perfect-match.html

Floridi, L. (2013). *The ethics of information.* Oxford University Press.
https://academic.oup.com/book/35378?login=false

Foley, S., Karlsen, J. R., & Putniņš, T. J. (2019). Sex, drugs, and Bitcoin: How much illegal activity is financed through cryptocurrencies? *The Review of Financial Studies, 32*(5), 1798-1853. https://www.researchgate.net/publication/333388187

García, R., Cediel, A., Teixidó, M., & Gil, R. M. (2025). A review of media copyright management using blockchain technologies from the academic and business perspectives. *Information, 16*(2), 72. https://www.mdpi.com/2078-2489/16/2/72

GeeksforGeeks. (2023). *Comparison - Centralized, decentralized and distributed systems - GeeksforGeeks.* https://www.geeksforgeeks.org/comparison-centralized-decentralized-and-distributed-systems/

GeeksforGeeks. (2025). *Proof of authority.* https://www.geeksforgeeks.org/computer-networks/proof-of-authority-consensus/

Gillis, A. (2023). *What is Hyperledger? Everything you need to know.* Search CIO. https://www.techtarget.com/searchcio/definition/Hyperledger

Gorkhali, A., & Chowdhury, R. (2022). Blockchain and the evolving financial market: A literature review. *World Scientific Connect.* https://www.worldscientific.com/doi/10.1142/S242486222150024X

Gubbi, J., Buyya, R., Marusic, S., & Palaniswami, M. (2013). Internet of Things (IoT): A vision, architectural elements, and future directions. *Future Generation Computer Systems, 29*(7), 1645-1660. https://www.sciencedirect.com/science/article/abs/pii/S0167739X13000241

Hahn, F. (2025). *Guide to creating a great blockchain UX.* Cheesecake Labs. https://cheesecakelabs.com/blog/ux-in-blockchain-web-3/

Harari, Y. N. (2014). *Sapiens: A brief history of humankind.* HarperCollins.

Hofstede, G. (1980). *Culture's consequences: International differences in work-related values.* SAGE Publications.

Hofstede, G. (2001). *Culture's consequences: Comparing values, behaviors, institutions and organizations across nations.* SAGE Publications. https://us.sagepub.com/en-us/nam/cultures-consequences/book9710

Hovard, P. (2023). *The Internet of Things: A landmark technology for behavior change?* Behavioral Economics. https://www.behavioraleconomics.com/the-internet-of-things-a-landmark-technology-for-behavior-change/

Hsieh, Y. Y., Vergne, J. P., Anderson, C. L., Lakhani, K. R., & Reitzig, M. (2017). The rise of the Bitcoin and the rise of decentralized autonomous

organizations. *Journal of Organization Design*, *6*(1), 1-14.
https://www.researchgate.net/publication/326816653

Huang, J., Zhou, K., Xiong, A., & Li, D. (2023). Smart contract vulnerability detection model based on multi-task learning. *International Journal of Environmental Research and Public Health*, *20*(4), 3350.
https://pmc.ncbi.nlm.nih.gov/articles/PMC8914670/

Hyperledger. (n.d.). *Cello*.
https://www.hyperledger.org/use/cello

Iaccarino, L. (2023). *Exploring the Dapps ecosystem: Empirical analysis of decentralized applications* [Master's thesis, Politecnico di Milano]. POLITesi.
https://www.politesi.polimi.it/retrieve/ad60c573-4d62-4392-b510-4c9e807c402f/Lorenzo_Iaccarino_Master_of_Science_Thesis_Final_Draft.pdf

ImmuneBytes. (2023, March 1). *An insight into the DAO attack*. https://immunebytes.com/blog/an-insight-into-the-dao-attack/

ImmuneBytes. (2024). *DAO governance attacks and how to prevent them - QuillAudits*. QuillAudits.
https://www.quillaudits.com/blog/web3-security/dao-governance-attacks

Kahneman, D. (2011). *Thinking, fast and slow*. Farrar, Straus and Giroux.

Kahneman, D., Knetsch, J. L., & Thaler, R. H. (1991). Anomalies: The endowment effect, loss aversion, and status quo bias. *Journal of Economic Perspectives, 5*(1), 193-206. https://www.aeaweb.org/articles?id=10.1257/jep.5.1.193

Kahneman, D., Sibony, O., & Sunstein, C. R. (2021). *Noise: A flaw in human judgment.* Little, Brown Spark.

Kahneman, D., & Tversky, A. (1974). Judgment under uncertainty: Heuristics and biases. *Science, 185*(4157), 1124–1131. https://sites.socsci.uci.edu/~bskyrms/bio/readings/tversky_k_heuristics_biases.pdf

Kanga. (2023). *How to avoid loss aversion in the crypto world.* https://kanga.exchange/loss-aversion

Kan, E. (2024). Block-chain and AI in healthcare data security: Creating a secure medical ecosystem. *International Journal of Law and Policy, 2,* 13-21. https://www.researchgate.net/publication/387546588

Karim, M. M., Van, D. H., Khan, S., Qu, Q., & Kholodov, Y. (2025). AI agents meet blockchain: A survey on secure and scalable collaboration for multi-agents. *Future Internet, 17*(2), 57. https://doi.org/10.3390/fi17020057

Kiely, E., & Robertson, L. (2020, May 11). *How to spot fake news.* FactCheck.org.

https://www.factcheck.org/2016/11/how-to-spot-fake-news/

Kotey, S. D., Tchao, E. T., Agbemenu, A. S., Ahmed, A. R., & Keelson, E. (2024). A framework for full decentralization in blockchain interoperability. *International Journal of Advanced Engineering Research and Science, 11*(4). https://pmc.ncbi.nlm.nih.gov/articles/PMC11645007/

Kowalski, P., & Esposito, L. (2023). The role of blockchain technology in enhancing supply chain transparency in Europe. *Journal of Procurement & Supply Chain, 7*(2), 11-21. https://www.stratfordjournals.com/journals/index.php/journal-of-procurement-supply/article/download/1694/2204/5372

Kshetri, N. (2021). Blockchain as a tool to facilitate property rights protection in the Global South: lessons from India's Andhra Pradesh state. *Third World Quarterly, 43*, 1-22. https://www.researchgate.net/publication/357144447

Lamport, L., Shostak, R., & Pease, M. (1982). The Byzantine generals problem. *ACM Transactions on Programming Languages and Systems (TOPLAS), 4*(3), 382-401. https://dl.acm.org/doi/10.1145/357172.357176

Lango, L. (2021). *The 5 phases of the crypto hype cycle (beware the big crash)*. Nasdaq. https://www.nasdaq.com/articles/the-5-phases-of-the-crypto-hype-cycle-beware-the-big-crash-2021-05-05

Lee, S., Trimi, S., & Kim, C. (2013). The impact of cultural differences on technology adoption. *Technological Forecasting and Social Change, 80*(4), 724-733. https://www.sciencedirect.com/science/article/abs/pii/S1090951612000405

Maharjan, P. (2018). *Performance analysis of blockchain platforms* [Master's thesis, University of Nevada, Las Vegas]. UNLV Theses, Dissertations, Professional Papers, and Capstones. http://dx.doi.org/10.34917/14139888

McLeod, S. (n.d.). *Making sense of blockchain technology | Karbon resources*. Karbon. https://karbonhq.com/resources/making-sense-of-blockchain-technology/

Mehrabi, N., Morstatter, F., Saxena, N., Lerman, K., & Galstyan, A. (2019). *A survey on bias and fairness in machine learning*. arXiv. https://arxiv.org/abs/1908.09635

Meiklejohn, S., Pomarole, M., Jordan, G., Levchenko, K., McCoy, D., Voelker, G., & Savage, S. (2016). A fistful of Bitcoins: characterizing payments among men with no names. *Communications of the ACM, 59*(4), 86–93. https://doi.org/10.1145/2896384

Merkle, R. C. (1980). Protocols for public key cryptosystems. *IEEE Symposium on Security and Privacy*, 122-133. https://www.ralphmerkle.com/papers/Protocols.pdf

Mhlanga, D. (2022). Block chain technology for digital financial inclusion in the industry 4.0, towards sustainable development? *SSRN Electronic Journal.* https://www.researchgate.net/publication/364409716

Minhaz, M. F., Sunny, M. S. H., Islam, M. M., & Nandi, D. (2024). *Enabling technologies for Web 3.0: A comprehensive survey.* arXiv. https://arxiv.org/html/2401.10901v1

Mitchell, P. I. (n.d.). *Changes in medieval commerce and production.* DBU.edu. https://www.dbu.edu/mitchell/medieval-resources/economicchangesmedieval.html

Mukherjee, S., Tang, W., Aniceto, G., Chandler, J., Song, W., & Jung, T. (2025). *Web3DB: Web 3.0 RDBMS for individual data ownership.* ResearchGate. https://www.researchgate.net/publication/390468927

Nadler, A., Segev, S., & Gal, U. (2024). *DAOs of collective intelligence? Unraveling the complexity of blockchain governance in decentralized autonomous organizations.* arXiv. https://arxiv.org/html/2409.01823v1

Nakamoto, S. (2008). *Bitcoin: A peer-to-peer electronic cash system.* https://bitcoin.org/bitcoin.pdf

Narain, A., & Moretti, M. (2022). Regulating crypto. *International Monetary Fund (IMF) Finance & Development, 59*(3), 39-41. https://www.imf.org/en/Publications/fandd/issues/2022/09/Regulating-crypto-Narain-Moretti

Nelson, J., & Charlotte, J. (2025). *Blockchain in digital identity management: Enhancing privacy and user control*. ResearchGate. https://www.researchgate.net/publication/390833022

Nickerson, R. S. (1998). Confirmation bias: A ubiquitous phenomenon in many guises. *Review of General Psychology*, 2(2), 175-220. https://journals.sagepub.com/doi/10.1037/1089-2680.2.2.175

Nielsen, J. (1994). *Usability engineering*. Morgan Kaufmann. https://www.nngroup.com/books/usability-engineering/

Omar, A., Bhuiyan, M. A. S., Alrubaian, M., Hawrylak, P. J., Al-Hammadi, Y., & Aktas, M. (2025). Blockchain-assisted self-sovereign identities on education: A survey. *Blockchains*, 3(1), 3. https://www.mdpi.com/2813-5288/3/1/3

OSL. (2025a). *Blockchain consensus mechanisms: PoW and PoS*. https://osl.com/academy/article/blockchain-consensus-mechanisms-pow-and-pos

OSL. (2025b). *What is a 51% attack?* https://www.osl.com/hk-en/academy/article/what-is-a-51-attack

Park, K. S. (2024). Property and sovereignty in America: A history of title registries & jurisdictional power. *The Yale Law Journal*, 133(4). https://www.yalelawjournal.org/article/property-and-sovereignty-in-america-a-history-of-title-registries-jurisdictional-power

Patel, O. (2024). *AI-driven smart contracts.* ResearchGate. https://www.researchgate.net/publication/383605866

Paul, A., & Ogburie, C. (2025). *The role of AI in preventing financial fraud and enhancing compliance.* GSC Online Press. https://gsconlinepress.com/journals/gscarr/sites/default/files/GSCARR-2025-0086.pdf

Pömer, A. Blog. (2025). *Web 1.0, 2.0, 3.0 and 4.0: The evolution of the internet.* SEEBURGER. https://blog.seeburger.com/the-evolution-of-the-internet-web-1-0-web-2-0-web-3-0-web-4-0/

Reuter, C., Stieglitz, S., & Imran, M. (2020). Social media in conflicts and crises. *International Journal of Information Management, 50,* 201-205. https://www.tandfonline.com/doi/full/10.1080/0144929X.2019.1629025

Roberts, R., Millar, D., Corradi, L., & Flin, R. (2021). What use is technology if no one uses it? The psychological factors that influence technology adoption decisions in oil and gas. *Technology, Mind, and Behavior, 2*(1). https://doi.org/10.1037/tmb0000027

Rokeach, M. (1960). *The open and closed mind: Investigations into the nature of belief systems and personality systems.* Basic Books.

Roy, G. (2023). *What is delegated proof-of-stake (DPoS)?* Ledger. https://www.ledger.com/academy/what-is-delegated-proof-of-stake-dpos

Samson, F., & Williams, A. (2025). *Integrating AI-driven analytics and blockchain technology for enhanced business intelligence: A comprehe*[sic]. ResearchGate. https://www.researchgate.net/publication/389509327

Santana, D., & Albareda, L. (2024). Exploring decentralized autonomous organization (DAO) governance: An integrative literature review. *ResearchGate.* https://www.researchgate.net/publication/385694204

Schär, F. (2021). Decentralized finance: On blockchain- and smart contract-based financial markets. *Federal Reserve Bank of St. Louis Review, 103*(2), 153-174. https://ideas.repec.org/a/fip/fedlrv/91428.html

Schmitt, L. (2025). Blockchain-based solution for tracking mission-critical data in a value chain. *The Accounting Review.* https://harbert.auburn.edu/news/2025/02/blockchain-based-solution-for-tracking-mission-critical-data-within-value-chain.html

Schneider, J. (2024). *A brief history of cryptography: Sending secret messages throughout time.* IBM. https://www.ibm.com/think/topics/cryptography-history

Sharma, T. K. (2024). *Types of blockchains explained- public VS private VS consortium.* Blockchain Council. https://www.blockchain-council.org/blockchain/types-of-blockchains-explained-public-vs-private-vs-consortium/

Silverbreit, A. (2025). *Regulating Web 3.0 for a safer digital future.* The Regulatory Review. https://www.theregreview.org/2025/01/07/silverbreit-regulating-web-3-0-for-a-safer-digital-future/

Smajgl, A., & Schweik, C. (2022). Advancing sustainability with blockchain-based incentives and institutions. *ScholarWorks@UMass.* https://scholarworks.umass.edu/entities/publication/02f88390-c5d3-4062-beb3-fed1c24ae38f

SNS Insider. (2025). *Blockchain market size to surpass USD 988.83 billion by 2032 | SNS Insider.* GlobeNewswire. https://www.globenewswire.com/news-release/2025/02/19/3028897/0/en/Blockchain-Market-Size-to-Surpass-USD-988-83-Billion-by-2032-SNS-Insider.html

Spar, I. (2004). *The origins of writing.* The Metropolitan Museum of Art. https://www.metmuseum.org/essays/the-origins-of-writing

Stoll, C., Klaaßen, L., & Gallersdörfer, U. (2018). The carbon footprint of Bitcoin. *Joule, 3*(7), 1647-1661. https://ceepr.mit.edu/wp-content/uploads/2021/09/2018-018.pdf

Sun, H., Wang, Y., & Li, X. (2025). *From data behavior to code analysis: A multimodal study on security and privacy challenges in blockchain-based DApp.* arXiv. https://arxiv.org/html/2504.11860v1

Sunstein, C. R. (2016). The ethics of nudging. *Behavioral Science & Policy, 2*(1), 6-12.

Thaler, R. H. (1985). Mental accounting and consumer choice. *Marketing Science, 4*(3), 199-214. https://www.semanticscholar.org/paper/Mental-Accounting-and-Consumer-Choice-Thaler/2d28a52e59005b2d3e23c9366be880a960dc1eaf

Thaler, R. H. (2015). *Misbehaving: The making of behavioral economics.* W. W. Norton & Company.

Thaler, R. H., & Sunstein, C. R. (2008). *Nudge: Improving decisions about health, wealth, and happiness.* Yale University Press.

Trigsted, M. (2025, February 14). *Blockchain identity management: A complete guide.* 1Kosmos. https://www.1kosmos.com/blockchain/blockchain-identity-management-a-complete-guide/

Tversky, A., & Kahneman, D. (1973). Availability: A heuristic for judging frequency and probability. *Cognitive Psychology, 5*(2), 207-232. https://www.sciencedirect.com/science/article/abs/pii/0010028573900339

Vendette, S., & Thundiyil, T. G. (2023). View of decentralization in blockchain: Reconsidering change

management theories. *Article Gateway, 23*(3), 102-110.
https://articlegateway.com/index.php/AJM/article/view/6376/6021

Wei, Z., Mo, S., Wang, B., Ding, K., & Long, J. (2024). *Token incentives in blockchain-based decentralized autonomous organizations* [Doctoral dissertation, City University of Hong Kong]. CityUHK Scholars.
https://scholars.cityu.edu.hk/en/theses/token-incentives-in-blockchainbased-decentralized-autonomous-organizations(8069232e-3de8-446f-b6cf-946adab46a8e).html

Weinberg, J. (2013). *The great recession and its aftermath*. Federal Reserve History.
https://www.federalreservehistory.org/essays/great-recession-and-its-aftermath

Wijesekara, P. A. D. S. (2025). An overview on blockchain-based social media. *Science Engineering and Technology, 5*(1), 222.
https://www.researchgate.net/publication/390027715

Wikipedia. (n.d.). *Bitcoin Foundation*.
https://en.wikipedia.org/wiki/Bitcoin_Foundation

Xu, J., & Livshits, B. (2018). The anatomy of a cryptocurrency pump-and-dump scheme. *Proceedings of the 27th USENIX Security Symposium (USENIX Security 18)*, 1609-1625.
https://www.usenix.org/system/files/sec19-xu-jiahua_0.pdf

Zhou, Q., Huang, H., & Zheng, Z. (2023). *Overview to blockchain scalability challenges and solutions.* ResearchGate.
https://www.researchgate.net/publication/371851330

About the Author

C.J. Carswell, a behavioral scientist, is deeply passionate about societal progress and empowering others through technology. Her journey into the world of FinTech began with cryptocurrency investing, sparking a deep interest in blockchain technology and its potential to revolutionize business operations and foster more collaborative online communities.

Carswell's research focuses on the human side of blockchain, exploring its impact on business operations, strategic management, and the development of more collaborative and generous online spaces. Through her work, she examines the psychological factors that influence the adoption and effectiveness of decentralized technologies.

Published Articles

- Decoding the Web: From Humble Beginnings to a Blockchain Future (Part 1)

- Decoding the Web: The Human Element of Blockchain

- Decoding the Web: The Human Side of Blockchain-Utopianism

- The Human Side of Blockchain-Egalitarianism

White Paper

BE DAO: A White Paper for Managing Blockchain Communities - A detailed blueprint on how to incorporate behavioral economics into DAO governance.

Contact Information

For inquiries, collaborations, or speaking engagements, please contact info@blocpsych.com.

www.ingramcontent.com/pod-product-compliance
Lightning Source LLC
Chambersburg PA
CBHW052003270326
41929CB00015B/2773